OECD Regional Development Studies

Reaching Climate Neutrality for the Hamburg Economy by 2040

OECD
BETTER POLICIES FOR BETTER LIVES

This document, as well as any data and map included herein, are without prejudice to the status of or sovereignty over any territory, to the delimitation of international frontiers and boundaries and to the name of any territory, city or area.

Please cite this publication as:
OECD (2024), *Reaching Climate Neutrality for the Hamburg Economy by 2040*, OECD Regional Development Studies, OECD Publishing, Paris, https://doi.org/10.1787/e1e44672-en.

ISBN 978-92-64-77202-1 (print)
ISBN 978-92-64-61324-9 (pdf)
ISBN 978-92-64-71973-6 (HTML)
ISBN 978-92-64-48506-8 (epub)

OECD Regional Development Studies
ISSN 2789-7990 (print)
ISSN 2789-8008 (online)

Photo credits: Cover © Hamburg Mediaserver/aufwind-luftbilder.photoshelter.com

Corrigenda to OECD publications may be found on line at: www.oecd.org/about/publishing/corrigenda.htm.

Foreword

Tackling the climate emergency requires economic transformations of unprecedented scale and speed. But while the OECD's International Programme for Action on Climate has registered 105 countries with carbon neutrality targets, often by 2050, many are off-track to meet those targets. Indeed, at the global level, climate actions are currently falling short of the measures needed to match the Paris Agreement goals.

This report aims at accelerating climate action to close this gap. It has its origin in the target, declared in 2021 by the Hamburg Chamber of Commerce (HCC) for the business community it represents, to reach climate neutrality by 2040. This is 5 years ahead of the Hamburg city region's and Germany's climate neutrality targets, and 10 years ahead of the European Union's. The HCC target and this report, prepared with the support of HCC, testify to the commitment of the Hamburg business community to lead by example.

This report addresses the challenge from a business and a regional perspective.

- A business perspective means defining concrete action towards making business models consistent with climate neutrality. With HCC's ambitious climate target, Hamburg businesses will need to get ahead of the policy agenda to gear business models to climate neutrality, avoiding failed investments and identifying related challenges and opportunities.
- As this report illustrates, embedding climate action in local economic contexts goes a long way to making climate neutrality an economic success. One reason is individual business action is not enough. Businesses need to work together, often across sectors, to create local synergies and to better exploit new shared infrastructures and ways of transforming and using energy. In Hamburg, this includes making the most of the opportunities provided by Europe's third biggest port, which is an integral part of Hamburg's ecosystem.

Taking a regional and business perspective reinforces policy action at the national and international level. This permeates all four chapters of the report.

The report was discussed at the Regional Development Policy Committee (RDPC) on 17 November 2023 and was approved by the RDPC through written procedure on 28 November 2023.

Acknowledgements

This report was produced by the OECD Centre for Entrepreneurship, SMEs, Regions and Cities (CFE), led by Lamia Kamal-Chaoui, Director, as part of the Programme of Work of the Regional Development Policy Committee (RDPC). It was produced with financial support from the Hamburg Chamber of Commerce (HCC).

The report was co-ordinated and co-authored by Andrés Fuentes Hutfilter, Head of Unit, under the supervision of Rüdiger Ahrend, Head of the Economic Analysis, Statistics and Data Division of CFE. Further co-authors are Ha-Young Jeon (CFE) and Matteo Schleicher (CFE) (Chapter 1), Caroline Gentet-Raskopf (CFE), Anna Molchanova (HCC), Matteo Schleicher and Valentina Ventricelli (CFE) (Chapter 2), Caroline Gentet-Raskopf, Ronald Halim (Equitable Maritime Consulting), Geneviève McInnes (consultant), Anaïs Rault (CFE) and Ha-Young Jeon (Chapter 3), Ander Eizaguirre (CFE), Mariam Fofana (CFE) and Oriana Romano (Chapter 4). Anna Molchanova and Simone Ruschmann (both Department for Climate Action, Energy and Manufacturing Industries, HCC) provided statistical resources and advice. The report was prepared for publication by Pilar Philip. Carlos Lerma-Poveda provided technical and editing support.

The OECD would like to thank Malte Heyne, Dirk Lau and Jan-Oliver Siebrand (all HCC) for their strong support throughout the preparation of the report as well as for useful comments.

The authors also wish to thank Sebastian Averdung (Council for the Energy Transition, Hamburg), Stefan Aykut (University of Hamburg), Lutz Birke (Ministry for Economic Affairs, City of Hamburg), Claudia Bühler (Michael Otto Foundation), Henner Buhck (Energy and Environment Committee, HCC), Timo Busch (University of Hamburg), Kai Gerullis (HCC), Benno Hain (Ministry for Environment, Climate and Agriculture, City of Hamburg), Daniela Jacob (Hamburg Climate Council), Heinz Lehmann (Urban Development Committee, HCC), Kerstin Kuchta (Technical University Hamburg), Marcus Merkenich (Chancellery, City of Hamburg) and Kerstin Wendt-Heinrich (Logistics, Port and Shipping Committee, HCC). The report also benefitted from valuable inputs from Matthias Ederhof (EnergieNetz-Hamburg) and Nick Zippel (Sager & Deus and Cpländer Haustechnik) as well as numerous further interview partners from the Hamburg business community, the City government and research institutions. The authors are also indebted to Nadim Ahmad, Soo-Jin Kim, Sandrine Kergroach, Oriana Romano (all OECD/CFE), Peter Levi (International Energy Agency) and other colleagues for their comments. The OECD Secretariat also thanks the delegates to the RDPC for their valuable comments.

Table of contents

FIGURES

TABLES

BOXES

Follow OECD Publications on:

http://twitter.com/OECD_Pubs

http://www.facebook.com/OECDPublications

http://www.linkedin.com/groups/OECD-Publications-4645871

http://www.youtube.com/oecdilibrary

OECD Alerts http://www.oecd.org/oecddirect/

This book has...

StatLinks
A service that delivers Excel® files from the printed page!

Look for the *StatLink* at the bottom of the tables or graphs in this book. To download the matching Excel® spreadsheet, just type the link into your Internet browser or click on the link from the digital version.

Acronyms and abbreviations

ACEA	European Automobile Manufacturers Association
ALS	Area Licensing Scheme
B2B	Business-to-business
BECCS	Bioenergy use combined with carbon capture and storage
BEVs	Battery electric vehicles
BMDV	Federal Ministry of Digital Affairs and Transport
BMUV	Federal Ministry for the Environment, Nature Conservation, Nuclear Safety and Consumer Protection (*Bundesministerium für Umwelt, Naturschutz, nukleare Sicherheit und Verbraucherschutz*)
BMWK	Federal Ministry for Economic Affairs and Climate Action (*Bundesministerium für Wirtschaft und Klimaschutz*)
BNW	Association for Sustainable Economy
BOF	Basic oxygen furnace
BSW	*Behörde für Stadtentwicklung und Wohnen*
BUKEA	Ministry of Environment, Climate, Energy and Agriculture of the City of Hamburg
CAPEX	Capital expenditures
CCS	Carbon capture and storage
CCU	Carbon capture and use
CCUS	Carbon capture, use and storage
CDR	Carbon dioxide removal
CII	Carbon intensity indicator
CMF	Clean marine fuels
CO2	Carbon dioxide
CSRD	Corporate Sustainability Reporting Directive
DAC	Direct air capture
DBU	German Federal Foundation for the Environment
DCCAE	Department of Communications, Climate Action and Environment of Ireland
DG-ECFIN	Directorate-General Economic and Financial Affairs of the European Commission
Difu	German Institute of Urban Affairs (*Deutsches Institut für Urbanistik*)
EAA	European Aluminium Association
EC	European Commission
ECI	European Copper Institute
ECJRC	European Commission's Joint Research Centre
EDGAR	Emissions Database for Global Atmospheric Research
EEDI	Energy Efficiency Design Index
EEXI	Energy Efficiency Existing Ship Index
EIT	European Institute of Innovation and Technology
EMC	Equitable Maritime Company
EPA	Irish Environmental Protection Agency
ESI	Environmental Ship Index
EU	European Union
EU CBAM	European Union Carbon Border Adjustment Mechanism
EU ETS	European Union Emissions Trading System
EUR	Euro
EV	Electric vehicle

EWKVerbotsV	Single-Use Plastics Labelling Ordinance (*Einwegkunststoffverbotsverordnung*)
FAO	Food and Agriculture Organization
FUA	Functional Urban Area
FCEVs	Hydrogen-powered fuel cell electric vehicles
GDP	Gross domestic product
GFS	Global Fuels Standards
GHG	Greenhouse gas
GLAD	Global LCA Data Access
GREET	Greenhouse gases, regulated emissions, and energy use in transportation
GT	Giga tons
GVC	Global value chain
GW	Gigawatt
H2	Hydrogen
HCC	Hamburg Chamber of Commerce
HDE	German Retail Association (*Kreislaufwirtschaftsgesetz*)
HEI	High Education Institution
HiiCCE	Institute for Innovation, Climate Protection, and Circular Economy
HmbKliSchG	Hamburg Climate Protection Act
HYBRIT	Hydrogen Breakthrough Ironmaking Technology
ICAO	International Civil Aviation Organization
IEA	International Energy Agency
IFB Hamburg	Hamburg Investment and Development Bank (*Hamburgische Investitions- und Förderbank*)
IFM	International Freight Model
IISI	International Iron and Steel Institute
IMO	International Maritime Organisation
IMR	Irish Manufacturing Research
IPCC	Intergovernmental Panel on Climate Change
ISO	International Organisation for Standardisation
ISWG GHG	Intersessional Working Group on GHG Emissions
ITF	International Transport Forum
ITPCO	International Taskforce on Port Call Optimisation
IAPH	International Association of Ports and Harbours
ICT	Information and communications technology
IT	Information Technology
KIBS	Knowledge-intensive business services
KIC	Knowledge and Innovation Community
KrWG	Closed Substance Cycle Management Act (*Kreislaufwirtschaftsgesetz*)
KSO	Climate Protection Initiative (*Klimaschutzoffensive*)
kWp	Kilowatt peak
LDCs	Least Developed Countries
LED	Light-emitting diode
LH2	Liquified hydrogen
LIHH	Logistics Initiative Hamburg
LPG	Liquified petroleum gas
LRT	Light-rail transit
LULUCF	Land use, land-use change and forestry
MEPC	Marine Environmental Protection Committee
MWp	Megawatt peak
NEA	Nuclear Energy Agency
NEW 4.0	North German Energy Transition initiative
NH3	Ammonia
NOW	*Nationale Online-Weiterbildungsplattform*
NOx	Nitric oxide
NZT	Net-zero target
OPEX	Operational expenditures

OVAM	Flemish Public Waste Agency
P2P	Peer-to-peer
PAW	Natural Gas-Free Neighbourhood programme
PM	Particulate matter
PPP	Public-private partnership
PRI	Public Research Institutions
PV	Photovoltaic
R&D	Research and development
RCP	Representative Concentration Pathway
RECE-XG	Expert Group on a new generation of information for a resource-efficient and circular economy
RISN	Resource Innovation and Solutions Network
SAN	Styrene Acrylonitrile
SBTi	Science-Based Target initiative
SDG	Sustainable Development Goal
SEAI	Sustainable Energy Authority of Ireland
SEK	Swedish Krona
SME	Small and medium-sized enterprise
SPW	Walloon Public Service
SRH	Hamburg City Cleaning Service (*Stadtreinigung Hamburg*)
SSP	Shared Socio-Economic Pathway
TCFD	Task Force on Climate-Related Financial Disclosures
TUHH	Hamburg University of Technology (*Technische Universität Hamburg*)
TWh	Terawatt-hour
UAE	United Arab Emirates
UaW	United Against Waste Association
UfR	Companies for resource conservation (*Unternehmen für Ressourcenschutz*)
UK	United Kingdom
UN	United Nations
UNECE	United Nations Economic Commission for Europe
USA	United States of America
USD	United States Dollar
VC	Venture capitalist
VerpackG	Packaging Act (*Verpackungsgesetz*)
VRE	Variable energy use
WHO	World Health Organization
WTW	Well-to-wake

Executive summary

With the decision by the Hamburg Chamber of Commerce (HCC) to reach climate neutrality by 2040, Hamburg's business community has decided to tackle the climate challenge head-on, not least because future-proofing business plans and investments that are consistent with climate neutrality will help seize opportunities and save substantial costs. This report shows what reaching climate neutrality means for Hamburg businesses and identifies key actions they need to undertake.

Operationalising climate neutrality for Hamburg businesses

Reaching the 2040 climate neutrality target will require reaching net zero for direct (Scope 1) and indirect (Scope 2) emissions by Hamburg businesses from the use of electricity and heat in Hamburg. Whilst indirect emissions arising throughout value chains (Scope 3) could reach net zero later, following science-based emission reduction scenarios consistent with the Paris Agreement, businesses will also need to take them into account to become consistent with climate neutrality.t

Accounting for almost 50% of Hamburg's economic activity, Hamburg's small and medium-sized enterprises (SMEs) are also important contributors to Hamburg's greenhouse gas emissions. However, especially compared to larger firms, they often lack the resources to make and indeed, master, the transformations climate neutrality requires. Financing stands out as a key obstacle in the path of SMEs towards decarbonisation. Networks can play a crucial role in overcoming this barrier, for example in connecting SMEs with specialised investors. Disclosing net-zero targets, action plans and progress reports can also help businesses demonstrate they are on track and help secure access to green financing and financing from banks interested in delivering on their own net-zero targets, whilst also increasing the potential to integrate into value chains of larger enterprises. Leveraging business networks to pool resources and indeed, improve access to needed knowledge and technology for climate-neutral business models, are also important and cost-saving tools.

Businesses must ramp up investment in energy efficiency now

In commercial buildings, heat pumps are likely to become the main source of heat where district heating is not available. Indeed, with the lifespan of fossil fuel boilers typically lasting around 20-30 years, investments in new fossil fuel boilers are inconsistent with Hamburg's 2040 climate neutrality target or risk being cut short. At the same time, businesses also need to decarbonise buildings more generally and improve energy efficiency. Acting now to meet the 2040 targets, and anticipating tighter regulations, will allow businesses to avoid 'bunching' of demand for construction activity that will be needed, which risks aggravating skills shortages and raising costs. Accelerating the pipelines of skilled workers, notably in relation to the installation of photovoltaic panels, heat pumps and energy efficiency renovation, can also help address supply bottlenecks. Support programmes will need to make sure all businesses, in particular SMEs, are able to adapt to and meet new regulations, and new regulations need also to be sensitive to potentially onerous reporting burdens on SMEs, including those that may be fully compliant in practice.

Low-cost renewable energy, notably solar and wind power, presents an opportunity for businesses to reduce their energy costs. To increase the share of renewables in their energy mix, businesses need to adjust their electricity demand to the time profile of renewables whilst electrifying the bulk of their energy use. Businesses should invest in smart energy storage systems and tools to manage electricity demand working together to share related infrastructure. They should advocate flexible pricing over time and space. Even without such pricing, ramping up rooftop photovoltaic systems, enabling self-production and consumption of energy, is already profitable in Hamburg.

Decarbonising activity around Hamburg's port offers opportunities beyond the local economy

Hamburg's port has already taken a leading role by electrifying port operations and developing zero-emission fuel infrastructure for ships. Hamburg can position itself as a climate-neutral transport hub by 2040 helping businesses throughout western and central Europe. Hamburg stands out for the large share of rail in freight from and to the port. This gives Hamburg a competitive advantage in delivering zero-emission long-haul freight quickly and at a low cost. For remaining trips, electric trucks are moving fast to cost parity with diesel, requiring a rapid ramp-up of investment in such trucks and major charging infrastructure.

As a key European industrial port, Hamburg could become a significant hydrogen hub serving its own high local needs as well as those of neighbouring regions. Scaling up infrastructure to transport, store and process hydrogen can reduce costs. Hamburg has taken important steps in this direction. Green ammonia, produced from green hydrogen, may also play a major role in providing a cost-effective zero-emission fuel mix, and thereby avoid rising transport costs in the long-term, including for trade flows through the Hamburg port. This requires switching investment in ships and infrastructure accordingly in the near term.

The Hamburg port also hosts major basic metals and oil refining industries. Investment in industrial production assets needs to be net-zero consistent starting in 2025 to avoid stranded assets. Energy-saving equipment, access to competitive electricity and hydrogen are key for that. Also, new digital technologies, such as low-cost sensors or real-time tracking of materials composition, can support the needed transformations towards circular economy practices. The resulting reduction in raw material input is essential to climate-neutral value chains. The port's location can offer opportunities for business models for used materials' trade and transformation.

The circular economy as a driver for climate neutrality

A number of Hamburg-based companies have adopted circular business models based on product recycling and reuse, and offer maintenance, repair and refurbishment services. However, circular models remain marginal. Several obstacles hinder effective implementation: businesses are not sufficiently aware of the benefits of the circular economy and lack financial incentives, while regulation is not yet conducive to circular innovations. The HCC could raise awareness of circular business principles and promote circular business models to make production and consumption practices climate-neutral. Capacity building, including training programmes for the business community, could open sustainable business opportunities. The HCC could collaborate with the city to support market innovation through the creation of incubators, hubs and spaces for experimentation.

1 Making climate neutrality operational

Chapter 1 aims to make the 2040 climate neutrality target of the Hamburg Chamber of Commerce operational. This chapter starts with laying out what climate neutrality means for Hamburg businesses, providing the national and regional context of climate targets and drawing on international best practice recommendations. The chapter examines available data on greenhouse gas emissions, energy use and the sectoral composition of Hamburg's economy. Net-zero targets and action plans of other industrial port cities – Rotterdam, Stockholm and Seattle – with similar decarbonisation challenges offer valuable insights for identifying obstacles and opportunities in the transition to climate neutrality. Finally, the chapter highlights the potential co-benefits of local climate action for well-being and competitiveness, which can exceed the transition costs. It showcases how rapid progress in the decarbonisation of urban passenger transport can deliver such benefits.

The Hamburg Chamber of Commerce (HCC, Box 1.1) has set the target to reach climate neutrality by 2040 for its member businesses. This report aims to identify key actions the businesses in Hamburg need to undertake as well as key challenges they need to face to develop business models consistent with this climate neutrality objective. As the report highlights, building the transformations on the specific regional economic context is key to addressing challenges and opportunities and requires businesses to work together. This report will argue that businesses should anticipate policy action to be better prepared for the major transformations, to save unnecessary costs and be in a better position to address challenges and seize opportunities. While it is not directed at policymakers, some references will be made to how policy can support the Hamburg business community to best face their challenges and take advantage of opportunities.

This chapter will start off with making the target operational for the business community. The first section discusses how businesses should interpret the target. It needs to be coherent with the national and international targets that will in due course bind Hamburg businesses, notably climate neutrality targets in the EU, Germany and Hamburg.

The second section provides an overview of available data on Hamburg's greenhouse gas (GHG) emissions and economic activity as well as their sectoral composition, laying out how the city's economic sectors will be concerned.

The third section provides an overview of the climate targets and action plans in three selected, comparable cities in Europe and North America, Rotterdam, Stockholm and Seattle. These cities have similar climate targets and face similar challenges. Businesses in Hamburg can learn from climate actions undertaken in these cities. Their business communities can cooperate with Hamburg's to reach their climate objectives. These cities are also competitors: Businesses in Hamburg need to make sure they do not fall behind in identifying opportunities and challenges.

Climate action can generate important wellbeing co-benefits, such as reduced air pollution or traffic congestion. Co-benefits materialise more quickly than climate benefits. They often exceed the local costs of climate action. Since many of these co-benefits arise locally and require local action, they can contribute to making regions more attractive and competitive and be a powerful motivator of local climate action. This issue is picked up in the final section of this chapter and illustrated with the example of urban passenger transport.

For the purposes of this report, the objectives of reaching climate neutrality and reaching net-zero GHG emissions will generally be used interchangeably. Most global warming is caused by long-lived GHG emissions, notably CO_2. These will need to be brought to net zero to halt global warming, as it is their cumulation in the atmosphere that determines global warming. Short-lived emissions, notably methane, may not need to reach net zero but still need to be halved by 2050 worldwide to be able to limit global warming to 1.5 degrees (Intergovernmental Panel on Climate Change, 2018[1]). The Hamburg territory emits little methane emissions. Still, methane plays an important role in the value chains of some Hamburg businesses.

The resource cost of reaching net zero GHG emissions by 2050 in high-income countries with modest fossil-fuel extraction and processing may amount to up to 1-2% of GDP (UK Committee on Climate Change, 2019[2]) Costs are concentrated on the last 10 - 20% of emissions abatement. The resource cost refers to the net resources that need to be devoted to the transition, including investment (European Commission, 2018[3]; OECD, 2017[4]). The impact of resource costs on the competitiveness of sectors subject to international competition depends on who bears them. Such sectoral competitiveness impacts may be particularly relevant if climate policies proceed at unequal speed across countries or regions. For example, if taxpayers assume resource costs, competitiveness in sectors subject to international competition may be largely preserved, and resources would not need to be reallocated to other sectors or geographies. When such reallocation occurs, it could further impact the distribution of economic activity across regions.

Early action is important for the climate but also to avoid unnecessary economic costs of delayed action. They can be large. The costs of delaying action to reduce GHG to meet the target of 1.5°C may be USD 5 trillion per year worldwide or 7% of the annual world GDP (Sanderson and O'Neill, 2020[5]). For Germany, net mitigation costs have been estimated to increase by an average of 40% for each decade of delay (Council of Economic Advisers, 2014[6]). A major source of additional costs from delayed action are investment decisions, especially for long-lived capital goods, that are inconsistent with climate objectives and which therefore need to be written off before their economic end of life ("stranded assets") These risks are particularly large in capital-intensive and energy-intensive activities, such as manufacturing activities (OECD, 2017[4]; OECD, 2023[7]). Further costs from delayed action arise from higher adjustment and coordination costs. Higher costs result because later reductions will require faster expansion of new technologies, raising susceptibility to errors (Chapman, 2019[8]).

Box 1.1. The Hamburg Chamber of Commerce

The Hamburg Chamber of Commerce (HCC) has approximately 170,000 business members registered in the federal state of Hamburg. It is among Germany's biggest. It is the representative organisation of Hamburg's firms and acts as political lobbyist, mediator and advocate for the local business community. In Germany all companies are required by law to be a member of a local chamber of commerce. The HCC is based on the participation and engagement of entrepreneurs in the region and acts autonomously, with its own individual responsibility as an organisation. The rationale for the mandatory membership of firms in the HCC is grounded in the need and wish of governments, parliaments and administrations to have a single point of contact with the local business and economic sectors.

As a lobbyist, the HCC is a strong advocate for a market-based legal and regulatory environment that is conducive to the business of small and medium-sized enterprises. Chambers of commerce are steered by elected entrepreneurs and managers and supported by their own employed staff. The aim of such joint teams of elected business representatives and chamber management is to provide policy advice in the "common interest" of the local economy.

As an independent mediator, the HCC supports fair business practices by offering a range of dispute resolution and prevention services. The HCC also acts as a service provider to the business community by providing non-market services to new and established firms.

The HCC business community covers a wide range of service sectors, such as retail, information and communication, real estate and housing, financial services, and transport. It also includes significant industrial businesses. Hamburg is the third largest civil aircraft construction city in the world and has a highly diverse industrial sector – there are few places in Europe with such a concentration of manufacturing industry. Some activities, in the crafts and in professional services, are covered by sector-specific chambers and are therefore not included in the chamber of commerce.

In 2020, the HCC launched the "Hamburg 2040" target vision and with it, the target to make Hamburg's business sector climate-neutral by 2040. The HCC has initiated a far-reaching dialogue on the question "In what world do we want to live in the future?" (*Wie wollen wir künftig leben – und wovon?*)

Figure 1.1. Hamburg 2040 target vision

Source: (Hamburg Chamber of Commerce, 2022[9])

Climate neutrality for Hamburg businesses in national and international context

The Hamburg economy will need to contribute to reaching net zero GHG emission targets in the European Union (EU), Germany and in the region (*Land*) of Hamburg. All three jurisdictions have set targets with legal force. The legal force of climate policy commitments is increasingly taken into account in court decisions in Germany and elsewhere. Meeting the HCC's climate neutrality objective should therefore serve Hamburg businesses to meet these legally enshrined emission targets, in terms of ambition and scope. It should also take into account the strong integration of Hamburg into the global economy.

The HCC's climate neutrality target is somewhat more ambitious in timing than the region's, Germany's and the European Union's (EU). Intermediate objectives are important to ensure early action and thereby give credibility to the climate neutrality objectives. Germany and the EU share a mid-term target with Hamburg's regional government: to reduce GHG emissions by at least 60% compared to 1990 levels by 2030. The Hamburg government aims to reduce CO_2 emissions by 70% below 1990 levels by 2030. The HCC has not set an intermediate target for its businesses, on aggregate or sectorally. Hamburg and Germany have sectoral decarbonisation targets with corresponding action plans to achieve them, though they have different categories of sectors. Power supply and buildings are expected to decarbonise the most quickly by 2030, relative to 1990 (Table 1.1). Hamburg businesses will need to lead the transformations required to reach climate neutrality – knowing that climate action is, currently, still far behind. They should be implementing transformations more quickly than other sectors, such as private households in Hamburg, as well as more quickly than other businesses in Germany and the EU. The Hamburg Port Authority has already set the target of climate neutrality for 2040 in port operations.

Table 1.1. Legally binding climate neutrality targets in the Region of Hamburg, Germany and the European Union

	Hamburg	Germany	European Union
		General targets	
Final target	- **98%** reduction of CO_2 emissions **by 2045** (net-zero CO_2 emissions with carbon sinks)	- **Net-zero GHG emissions by 2045** and net-negative after 2050	- **Net-zero GHG emissions by 2050**
Intermediate targets	- **70%** reduction of CO_2 emissions below 1990 levels **by 2030**	- **55-65%** reduction of national greenhouse gas (GHG) emissions below 1990 levels **by 2030**, excluding land use, land-use change and forestry (LULUCF) - **88%** reduction of national GHG emissions below 1990 levels **by 2040**, excluding land use, land-use change and forestry (LULUCF)	- **At least 55%** reduction of economy-wide net domestic greenhouse gas (GHG) emissions below 1990 levels by 2030
		Sectoral targets	
Power supply	No target	**61-62%** reduction of CO_2 emissions below 1990 levels **by 2030**	Not available
Industry	**73%** reduction of CO_2 emissions below 1990 levels **by 2030**, and **-99% by 2045**	**49-51%** reduction of CO_2 emissions below 1990 levels **by 2030**	Not available
Buildings	**68%** reduction of CO_2 emissions below 1990 levels **by 2030** and **-94% by 2045** (for residential buildings)	**66-67%** reduction of CO_2 emissions below 1990 levels **by 2030** (for all buildings)	Not available
Trade, commerce, services and non-residential buildings	**75%** reduction of CO_2 emissions below 1990 levels **by 2030** and **-99% by 2045**	Not available	Not available

Transport	**53%** reduction of CO_2 emissions below 1990 levels **by 2030** and **-100% by 2045**	**40-42%** reduction of CO_2 emissions below 1990 levels **by 2030**	**90%** reduction of CO_2 emissions below 1990 levels **by 2050**
Agriculture	No target	**41-43%** reduction of CO_2 emissions below 1990 levels **by 2030**	Not available
Other	Not available	**87%** reduction of CO_2 emissions below 1990 levels **by 2030**	Not available

Source: (Hamburger Senat, 2022[10]); (Federal Ministry for the Environment, Nature Conservation, Nuclear Safety and Consumer Protection of Germany, 2016[11]); European Climate Law (Article 2(1); Article 4(1)); (European Commission, 2022[12])

The scope of reaching climate neutrality in the Hamburg business community

The emissions coverage of the EU and German targets provides critical information on what emissions the climate neutrality objective should cover for the Hamburg business community. Both the EU and Germany define the emission objectives in terms of Scope 1 emissions (Box 1.2). These include all direct GHG emissions generated within their geographical boundary. The Hamburg region's target includes emissions from energy end use, including both Scope 1 and Scope 2 emissions. It does not include emissions on Hamburg territory from energy transformation, notably from the generation of electricity and oil refining. Consistent with this approach, local emissions from heat and electricity generation are attributed to end-users, even where electricity and heat are generated outside city borders.

Table 1.2. Emissions coverage in Hamburg, Germany, and EU net-zero emissions targets

	Hamburg	Germany	European Union
Emissions scope	**Scope 1 and 2 CO_2 emissions from energy use in energy end-use sectors**, where scope 2 emissions include emissions from oil refining. The remaining GHG emissions (process CO_2 emissions, methane, nitrous oxide and fluorinated gases) are also to be reduced.	**Production-based emissions (scope 1)**	**Production-based emissions (scope 1)**
International offset reliance	Not mentioned in the emissions reduction target.	**Mentioned** in the Federal Climate Law without any specific plan announced.	**Excluded** from the emissions reduction target.
International aviation and shipping emissions	**Excluded** from the emissions reduction target.	**Excluded** from the emissions reduction target.	**Partially included** in the emissions reduction target (extra-EU flights starting from the EU).
Contribution of carbon sinks from land use, land-use change and forestry (LULUCF) as well as from carbon capture, use and storage (CCUS)	**Mentioned** in the Hamburg Climate Plan as possible measures.	**LULUCF carbon sinks included** in the emissions reduction target **but with a limited role**. **CCUS mentioned** as an important element in the transformation of the industrial sector to offset unavoidable residual emissions.	**LULUCF carbon sinks limited** to 225Mt, around 6% of 2019 emissions. A target of 310 $MtCO_2$ for 2030 (8% of 2019 emissions) may be agreed. **CCUS limited** to industrial process emissions.

Source: (Hamburger Senat, 2022[10]); First revision of the Hamburg Climate Plan (p.28; Annex 4, p.5; Annex 4, p.7); German Climate Action Plan 2050 (p.50; 14 & 83; 67); (IEA, 2022[13]); German Federal Climate Change Act (Section 3, Clause 2);

Box 1.2. Three scopes of greenhouse gas (GHG) emissions

GHG emissions of subnational geographic levels, such as cities and regions can be defined according to three different scopes (World Resource Institute, C40 Cities Climate Leadership Group and ICLEI – Local Governments for Sustainability USA, 2021[14]). The same holds for businesses and institutions. For a city, the three scopes of emissions refer to:

- **Scope 1** emissions are direct GHG emissions occurring within the city boundaries.
- **Scope 2** emissions are indirect emissions occurring as a consequence of local consumption of electricity, heat, steam and/or cooling. The City of Hamburg also includes emissions from oil refining in Scope 2, attributing them to the consumers of oil refining production. Scope 2 emissions are commonly estimated by a location-based method, multiplying the amount of electricity, heat or steam purchased by the power supply's average emissions factor. Alternatively, a market-based method calculates Scope 2 emissions based on specific electricity purchase contracts with producers.
- **Scope 3** emissions are all other indirect emissions because of activities taking place within city boundaries. Hence, it includes all the upstream and downstream emissions in the value chains of local activities occurring outside the boundary of the city.

Consumption-based emissions are a way to capture indirect emissions of local activities which include only upstream emissions. Consumption-based emissions of a city refer to GHG emissions from the consumption of all goods and services finally consumed by city residents, irrespective of their origin". Consumption-based emissions are strongly related to consumption spending and income levels.

Consumption-based emissions are of particular interest in high-income cities, where they are often much higher than Scope 1 or Scope 2 emissions. Identifying consumption-based emissions allows high-income cities to reduce emissions from the demand side, often at low cost, for example in the consumption of food, drink and other consumption goods even if they are not produced locally (OECD, 2021[15]).

At the level of an individual geographic or institutional unit, the three scopes of emissions are mutually exclusive. However, if emissions of several scopes are added up across geographies or institutions, double counting arises, as emissions of one scope may be emissions of another scope in another unit.

Source: (Chen et al., 2018[16]); (Wiedmann et al., 2020[17]); (World Resource Institute, C40 Cities Climate Leadership Group and ICLEI – Local Governments for Sustainability USA, 2021[14]),

The EU requires the net-zero GHG emissions objective to be reached on domestic emissions. The purchase of emission reductions outside the EU does not count towards this target. The EU will contribute to financing emission reductions outside its borders but doing so will be additional to its net zero objective. German legislation does not exclude contributions from international offsets to reach its targets, but the government's political intention appears to keep its role small.

The EU and Germany also limit the role of carbon dioxide removal (CDR) to contribute to emission reductions. CDR can contribute to emission reduction by absorbing emissions durably elsewhere than in the atmosphere. Two main avenues are through land use and land use change as well as through carbon capture, use and storage (CCUS). Worldwide, the share of emissions that could be offset with CDR may be less than 20% and should therefore be limited to offsetting residual emissions in hard-to-decarbonise activities (Buck et al., 2023[18]).

The International Maritime Organisation (IMO) has adopted an objective to net zero GHG emissions in international shipping by 2050, although the role of offsets is unclear (Chapter 3). They account for 3% of the total GHG emissions worldwide (International Maritime Organization, 2020[19]). With regards to international aviation, emissions of extra-EU flights departing from EU territory are included in the EU's climate objectives based on fuel purchased domestically (German Presidency of the Council of the European Union, 2020[20]). International flights within EU territory are included in the European Emissions Trading System (EU ETS) (European Commission, 2021[21]). However, Germany and Hamburg do not include international aviation in emissions reductions in order to better reflect the local climate policy impact.

Hamburg is a highly internationally connected economy. Hence, the 2040 climate neutrality objective of the HCC needs to be placed in the broader context of worldwide climate objectives. Most OECD countries have adopted net-zero emission targets by 2050, although some have limited net-zero targets to CO_2 emissions and have given international offsets and CDR a bigger role. China aims to achieve net zero GHG emissions before 2060. India aims to reach net zero by 2070.

Businesses will trade with partner countries that will also need to reach climate neutrality, albeit possibly later. Businesses will need to transform purchases of intermediate goods and services from trading partners to take into account progress towards climate neutrality which is necessary also in other countries. Moreover, policy action is likely to require that imported goods and services meet increasingly stringent requirements on the emissions generated in production. The European Union has taken the first steps in this direction with its carbon border adjustments and the regulation of deforestation-free imported products (Chapter 2). Since Hamburg is a major trading hub with a major international port, understanding the GHG emissions embedded in the goods and services involved in this trade is particularly important for businesses in Hamburg. This international context is essential to understand business opportunities and challenges they face in the transition to climate neutrality.

This discussion of emission targets has the following implications for making the HCC climate neutrality target operational:

The climate neutrality objective set by the HCC should include reaching net zero GHG emissions for all direct (Scope 1) and indirect emissions from the use of electricity and heat (Scope 2 emissions) of businesses at least on the territory of Hamburg by 2040. Scope 1 emissions of Hamburg businesses generated elsewhere in Germany or on EU territory should reach net zero by 2045 and 2050 respectively. With power supply largely determined by German or EU production, which will largely be decarbonized by 2040, and district heat supply in Hamburg, Hamburg businesses should purchase all electricity and heat from zero-emission sources by 2040.

- **The HCC and its businesses should not rely on major international offsets to reach climate neutrality objectives.** International emission offsets may be reasonable between 2040 and 2050 for those business Scope 1 emissions on Hamburg territory that Hamburg businesses may only bring to net zero by 2050 but not by 2040. This could be particularly relevant for difficult-to-decarbonise sectors, including emission-intensive manufacturing. Scope 1 emissions should reach net zero without any international offsets by 2050 at the latest.

Emission reductions through carbon sinks should play a minor role. The use of CCS should be limited to process emissions in industry. Offsetting Scope 1 emissions on a small scale could also include financing LULUCF carbon sinks in other EU countries, but these should broadly be limited to the EU LULUCF target contribution share.

Hamburg businesses should include indirect emissions in climate neutrality targets. Taking into account scope 3 emissions at the individual business level will allow to fully integrate opportunities and challenges from reaching climate neutrality in business models. but on different time scales for scope 3 emissions.

Some scope 3 emissions of Hamburg businesses could reach climate neutrality after 2040. Scope 3 emissions targets could be set in a differentiated way, depending on goods and services and whether value chains originate in Germany, the EU and other OECD countries. Value chains originating in the EU should reach net zero GHG emissions by 2050 without offsets at the latest. Elsewhere, they should broadly follow science-based worldwide emission reduction scenarios consistent with limiting global warming to 1.5 degrees with at least 50% probability. This would require reaching net zero CO_2 emissions in value chains while halving methane emissions by 2050.

- **The HCC could set intermediate emission reduction targets for 2030 for scope 1 and scope 2 emissions** and provide guidance on intermediate scope 3 emissions targets for businesses.
- **The HCC may prepare investment guidance** to avoid costs from delayed action. It could indicate the latest point of time when purchases of new fossil-fuel-using equipment should be avoided. For example, with an average useful life of cars of 15 years, purchases of new cars with internal combustion engines, if used in Hamburg for their entire useful life, should be avoided from 2025.

Meeting these recommendations would align climate action by the HCC and its members with recommendations from the High-level Expert Group of the United Nations (UN) on net-zero commitments of non-state entities (Box 1.3). The expert group provides ten science-based best practices recommendations for non-state entities' net-zero claims which Hamburg businesses should follow (United Nations' High-Level Expert Group, 2022[22]).

Box 1.3. Ten recommendations for net-zero pledges by non-state actors by the High-level expert group of the United Nations (UN)

The ten recommendations are the following. They are directed at subnational governments and businesses.

1. **Announce a public net-zero pledge which contains intermediary targets (2025, 2030, 2035)** and is in line with the IPCC scenarios to limit warming to 1.5°C. Any actor with the capacity to move faster should do so.
2. **Set net-zero targets within a year of making the pledge.** Targets should be short, medium and long-term absolute and relative emission reduction targets across the value chain. Targets must include scope 1, 2, and 3 emissions. Emissions embedded within fossil fuel reserves and land-use-related emissions should be accounted for separately.
3. **Use high-integrity voluntary carbon credits for beyond value chain mitigation** but not counted towards intermediate reduction targets. Non-state actors who choose to purchase voluntary carbon credits for permanent removals for residual emissions or annual unabated emissions beyond their net zero pathways must use high-quality carbon credits. A high-quality carbon credit should at minimum fit the criteria of additionality and permanence. Additionality means the mitigation activity would not have happened without the incentive created by the carbon credit revenues. Permanence means that any carbon sequestered or avoided will remain out of the atmosphere. Any credit transaction must be transparently reported, and associated claims must be understandable, consistent and verified.
4. **Create and publicly disclose a transition plan,** which is comprehensive and which sets out actions that will be undertaken to meet all targets. It should describe how governance and incentive structures, capital expenditure, research and development, skills and human resources will be aligned for a just transition. Transition plans should be updated every year and progress should be reported annually.
5. **Phase out fossil fuels and scale up renewable energy**. The transition away from fossil fuels must be just for all the affected communities, workers and consumers. The transition away from fossil fuels must be matched by a fully funded transition toward renewable energy.
6. **Align external policy and engagement efforts** to the goal of reaching net zero by 2050. This means lobbying for positive climate action and not against it. Businesses should publicly disclose their affiliations.
7. **Include people and nature in the just transition efforts**. Achieve operations and supply chains that avoid the conversion of natural ecosystems. Eliminate deforestation and peatland loss by 2025, and the conversion of other remaining natural ecosystems by 2030 from operations and supply chains.
8. **Increase transparency and accountability** by reporting GHG data, net-zero targets, plans and progress in a standardised, open format and via a public platform. Disclosures should be accurate and reliable. Businesses should seek independent evaluation of their annual progress reporting and disclosures.
9. **Invest in just transition efforts**, for example, all businesses with operations in developing countries should demonstrate how their net zero transition plans contribute to the economic development of the regions they are operating in.
10. **Regulators should develop and accelerate regulation** and standards in areas including net-zero pledges, transition plans and disclosure.

Source: (United Nations' High-Level Expert Group, 2022[22])

The sectoral structure of GHG emissions in Hamburg

Understanding the sectoral composition of Hamburg's GHG emissions is essential to reach the climate neutrality objective. The Hamburg Statistics Office produces statistics for Scope 1 and Scope 2 energy-related CO_2 emissions. Emissions in this section refer to 2019 to avoid the COVID-19 lockdown effects on emissions in 2020. Data on Scope 3 emissions are not available.

According to the Hamburg Statistics Office, total Scope 1 CO_2 emissions decreased from 12.7 million tonnes in 1990 to 10.4 million tonnes in 2021 (Figure 1.2). Energy transformation and transport are the sectors generating the most emissions. Coal-fired electricity generation rose from 2013 to 2019, but fell with the closure of one of two remaining coal-fired heat and power plants in 2021. Emissions in energy transformation include close to 1 million tons of CO_2 emissions from oil refining.

Figure 1.2. The energy transformation sector is the largest CO_2 emitter in Hamburg

Energy-related CO_2 production-based emissions, Scope 1 in 1990 and 2021

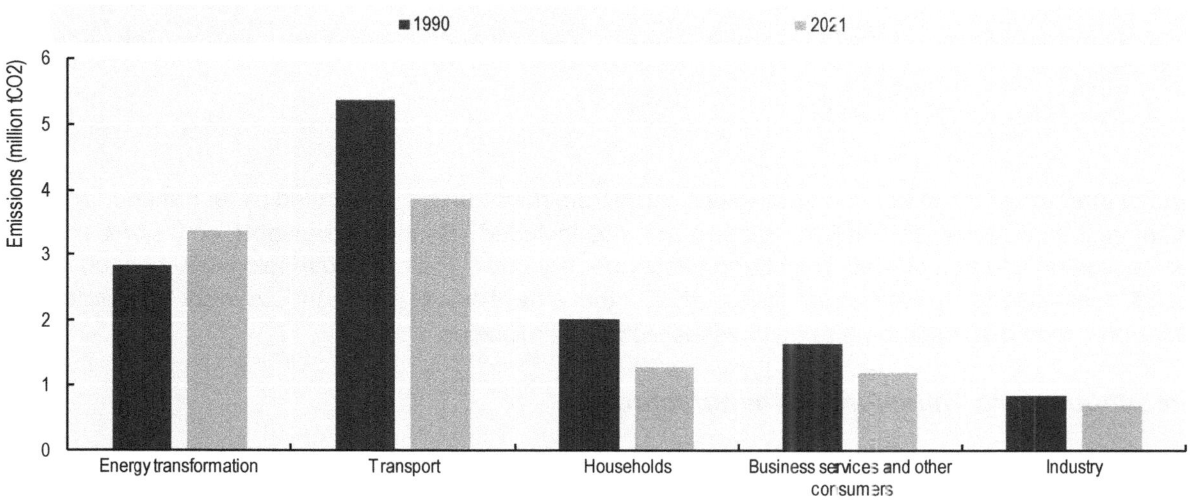

Note: Emissions that result from the use of primary energy sources ("Quellenbilanz"). Energy transformation includes electricity generation as well as oil refining. The transport sector includes domestic rail and road as well as air transport, and coastal and inland shipping. Other consumers include public facilities, small crafting businesses excluded from the industry sector, construction companies, and agriculture and forestry.
Source: (Hamburg Statistics Office, 2022[23]).

Hard coal accounted for approximately half of the energy transformation sector's Scope 1 emissions in 2021 (Figure 1.3). Emissions from energy transformation therefore fall more in 2022. Even so, natural gas contributed around a third to electricity generation (Figure 1.3). Natural gas also is a major emissions source in the industrial sector, in the other business sectors, and in households, where it is mostly used for heating and cooling of buildings. Heating oil also contributes to emissions in these sectors. The relatively small reductions in emissions between 1990 and 2021 indicate that action to reduce emissions needs to accelerate sharply to reach climate neutrality by 2040. Natural gas use and mineral oil use will need to be phased out by 2040. A conceivable exception is process emissions in manufacturing from gas or oil use, which could be abated with CCS although hydrogen use could avoid CCS, as discussed in chapter 3. Mineral oil use in energy transformation mostly reflects oil refining. Lignite and mineral oil in energy use will also need to be phased out by 2040.

Figure 1.3. Natural gas is the largest source of emissions in the business services sector

Scope 1 emissions by sectors and energy carriers, 2021

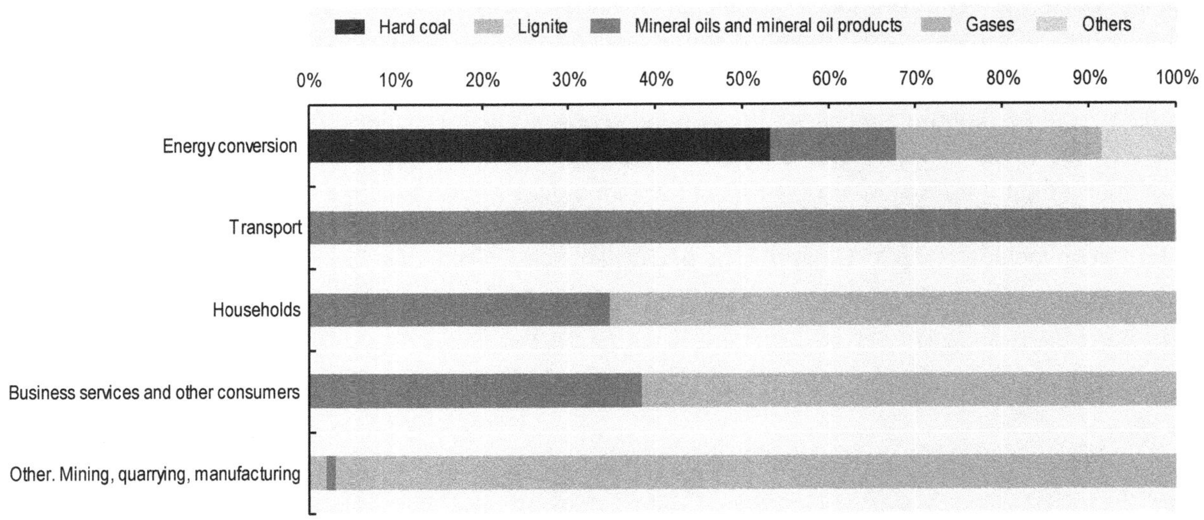

Source: (Hamburg Statistics Office, 2022[23]).

Around three-quarters of transport emissions come from road transport, followed by air transport and inland shipping. International shipping emissions are not included. Scope 1 transport emissions matter for businesses on account of road freight and passenger transport. Private households may account for most scope 1 emissions in passenger transport. To the extent travel from and to work generates these emissions, they also contribute to businesses' scope 3 emissions.

Greenhouse gas emissions in manufacturing

Manufacturing activities account for most industrial scope 1 emissions, as mining activity is minor. Emissions-intensive manufacturing of basic materials is among the most difficult to decarbonise (OECD, 2023[7]). Using EU ETS data allows for a sectoral breakdown, albeit with the limitation that small emitting installations may not be included (OECD, 2023[7]). Most manufacturing emissions in Hamburg arise in oil refining, followed by the production of iron and steel, and then the manufacturing of aluminium (Figure 1.4, Box 1.4) (OECD, 2023[7]) Industrial emissions presented by Hamburg Statistics in Figure 1.2 are lower, because they include oil refining in energy transformation rather than in industry.

Figure 1.4. Production of refined oil products, steel and aluminium are significant emitters

Emissions of Hamburg installations (2019)

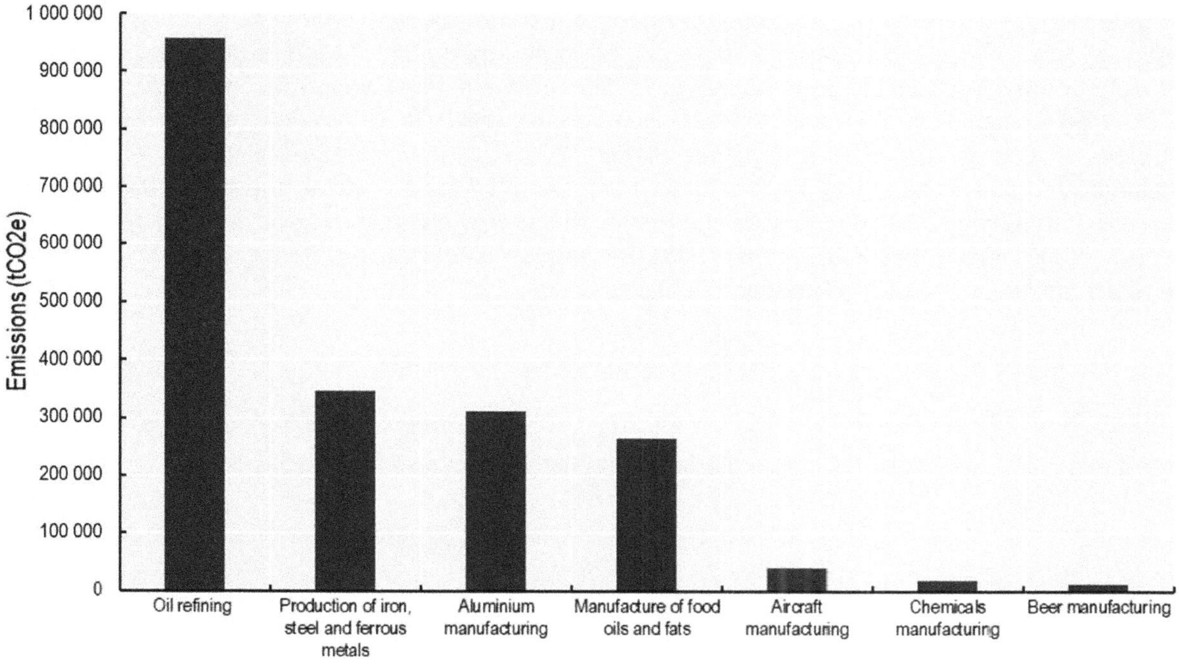

Note: EU ETS provides emissions data gathered from each factory and region. Combined with the firm-level installations data of the Orbis dataset, ETS data can provide emissions from the manufacturing sector of Hamburg.
Source: ETS-ORBIS matched dataset

Box 1.4. Basic material manufacturing activities in Hamburg

- **Oil refining** – the biggest company in this sector is Holborn, which produces mainly transport fuels. For climate neutrality, the company is focusing on a transition to e-fuels. They are synthetic fuels produced from renewable electricity, suitable for all modes of transport.
- **Steel** – the biggest actor in Hamburg is ArcelorMittal, one the world's leading steel and mining companies. In Hamburg, they produce billet and high-quality wire rode, mainly for automotive and engineering customers. In Hamburg, the company is collaborating with Hamburg's H2 project and is testing a carbon-free production of basic iron and steel.
- **Aluminium** – one of the companies producing primary aluminium in Hamburg is Trimet. Trimet aims to produce climate-neutral aluminium by 2045.
- **Copper** – Aurubis is the main copper producer and copper recycler in Hamburg. At the Hamburg site, the use of hydrogen was tested for the first time on an industrial scale in 2021. The plant in Hamburg has also been awarded the copper mark, the seal of quality for sustainability in the copper industry.

Adding indirect emissions from energy use raises industrial and service sector emissions

The Hamburg Statistics Office provides statistics that attribute emissions in energy transformation to energy end-use sectors by adding Scope 2 emissions to the Scope 1 emissions ("*Verursacherbilanz*"). Energy end-use sectors exclude activities engaged in energy transformation, notably the production of electricity and heat and oil refining. Indirect emissions from electricity generation are attributed following the average emissions in Germany. In 2021, Hamburg recorded 14.45 million tonnes of Scope 1 and 2 emissions in energy end-use sectors, 6.6 million tCO_2 less than in 1990. Scope 2 emissions add substantially to emissions in industry and other business activities as well as in households (Figure 1.5). This reflects the high share of electricity and heat in the energy mix of these sectors (Figure 1.6) and the relatively high share of emission-intensive coal in German electricity and heat generation. By contrast, Scope 2 adds only marginally to transport emissions.

Figure 1.5. Industrial activities are the second largest emitter when energy production emissions are counted

Scope 1 and 2 CO_2 emissions (Verursacherbilanz) by the energy end-use sector in 1990 and 2021

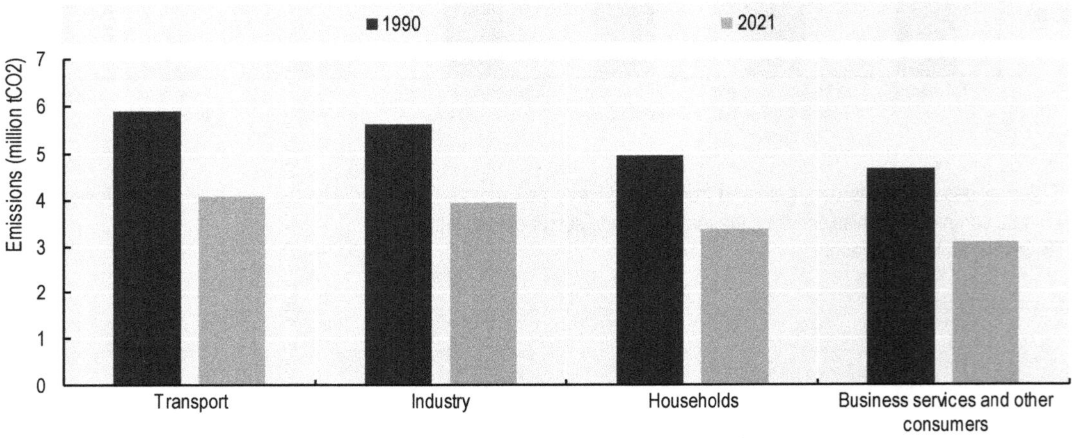

Note: The industry sector includes mining, quarrying, and manufacturing of materials and products. It does not include oil refining.
Source: (Hamburg Statistics Office, 2022[23]).

Figure 1.6. More than half of the Scope 2 emissions of the industry sector come from electricity generation

Share of Scope 2 emissions by energy sources in 2021

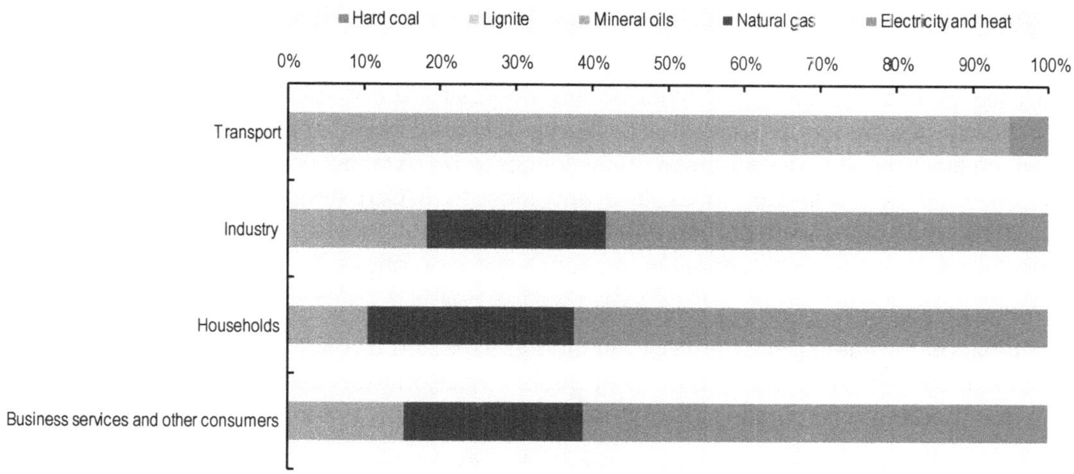

Source: (Hamburg Statistics Office, 2022[23]).

Germany-wide decarbonisation of electricity generation will reduce Scope 2 emissions, measured in this way, in Hamburg. As long as the energy mix is not fully decarbonised, Hamburg businesses can reduce their Scope 2 emissions by reducing energy use or by purchasing it from renewable sources. Businesses will not generally be able to choose a district heating provider. Climate-neutral district heating will require that heat generation from coal-fired power inside and outside Hamburg by the regional energy utility is phased out. The regional government, which owns the district heating provider, is committed to doing so by 2030.

A key challenge to reaching climate neutrality is the need to electrify most energy use while moving almost all electricity generation to renewables. According to the IEA, to reach net zero GHG emissions worldwide in 2050, electricity will need to represent 52% of the final energy consumption, a significant increase from the 20% in 2021. 88% of electricity generation will be from renewables (International Energy Agency, 2022[24]). In Europe, renewable generation is expected to increase by more than 380% to reach climate neutrality in 2050, compared to 2021 (International Energy Agency, 2022[24]). Germany aims to reach a share of renewables of 80% already in 2030. Without efforts to lower energy consumption, the needed expansion of renewable electricity production risks being at a scale that is difficult to manage. In a climate-neutral world, energy is therefore likely to be scarce. The energy intensity of GDP may need to fall by about two-thirds by 2050 worldwide (OECD, 2021[15]). In the EU, the RePowerEU plan sets an energy efficiency target of a 13% reduction in primary energy consumption (European Commission, 2023[25]). In addition to reducing costs, lowering energy demand also raises business resilience to energy supply shocks.

Some energy needs can be met from sustainable, zero-carbon sources without electrification. Biofuels can replace fossil fuels with relatively limited changes in equipment, such as in motor engines. Biofuels only eliminate emissions if sourced from the sustainable growth of biomass. However, the calls on bioenergy for industrial production alone are likely to exceed sustainable bioenergy supply (Material Economics, 2019[26]). Biomass growth also competes with essential land uses, notably for food production and biodiversity protection. Moreover, biomass supply is vulnerable to shocks, including from extreme climate events such as fires or drought.

"Green hydrogen", produced from renewable electricity and hydrogen-derived products, such as synthetic fuels, may also serve decarbonisation instead of electricity. This applies especially when electrification of energy use is difficult, for example, because temperatures in production processes are very high. However, transforming renewable electricity into hydrogen implies substantial energy loss. Green hydrogen production may be concentrated in regions across the globe with the highest renewables potential and will be internationally tradable and therefore subject to shocks that can be transmitted internationally.

Hence, the limitations of bioenergy and hydrogen, including hydrogen-derived fuels, argue to prioritise them for uses in which electricity or other energy sources are not suitable or insufficient. Priority use is notably in heavy-duty transport (air as well as heavy-duty road freight and ship) as well as in manufacturing sectors which are difficult to decarbonise. In the longer term, biomass firing is a key source of net negative emissions if combined with carbon capture and storage (CCS) (Intergovernmental Panel on Climate Change, 2019[27]). Relying on hydrogen and biomass when not necessary may also weaken resilience, as these energy sources may be susceptible to shocks and high prices.

Further renewable energy sources beyond electricity include solar and geothermal energy to produce heat. Their contribution to meeting final energy demand is expected to be limited. Globally, to meet projected energy consumption for net zero emissions by 2050, hydrogen may represent 6%, biofuels 4%, and heat, including from geothermal sources, 2% of the total energy supply (International Energy Agency, 2022[28]).

Transport uses the most energy across sectors in Hamburg (Figure 1.7), with road transport accounting for the largest share (Figure 1.8). Energy-saving transport modes are therefore particularly important. Transport is also the sector that is least advanced in the electrification of energy use (Figure 1.9). In industry and other business sectors electricity accounts for approximately half of energy use.

Figure 1.7. The transport sector consumes the most energy

Final energy consumption by emitting sectors in 2021, Hamburg

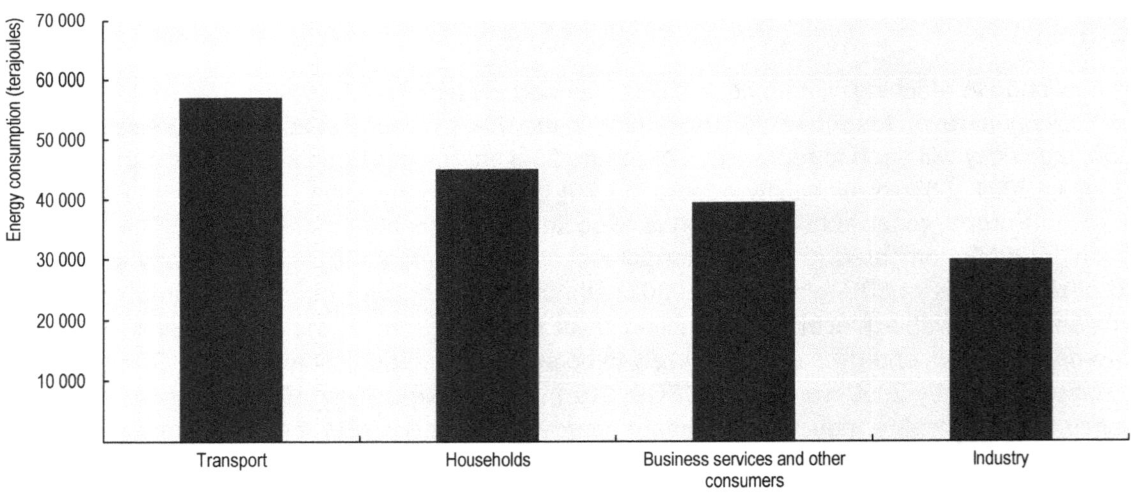

Source: (Hamburg Statistics Office, 2022[23]).

Figure 1.8. Road traffic is the biggest consumer of energy in the transport sector

Final Energy Consumption of the Transport Sector in 1990 and 2021 Hamburg

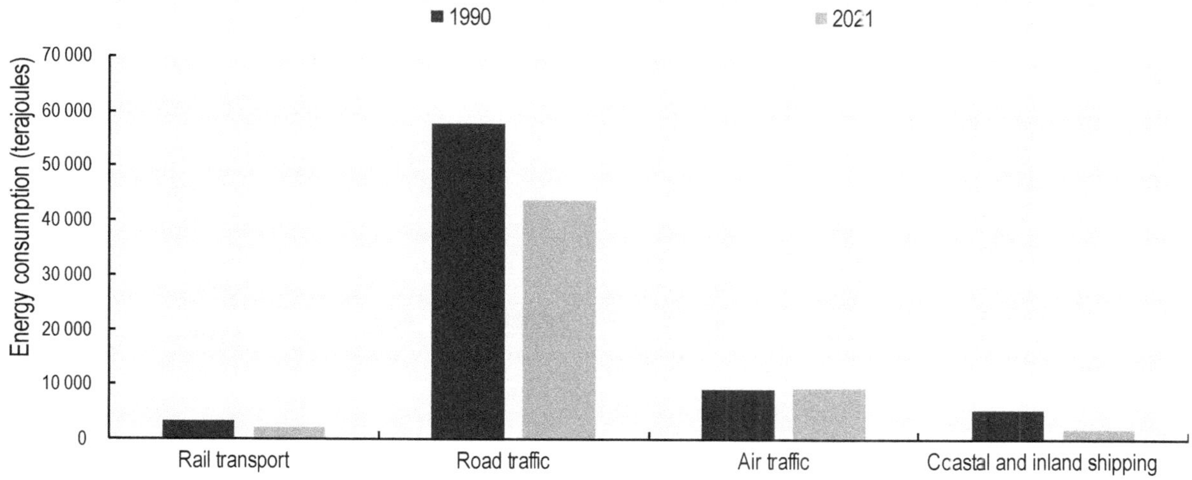

Source: (Hamburg Statistics Office, 2022[23]).

Figure 1.9. Fossil fuels account for at least 40% of energy end use in all sectors

Energy consumption share by sources in 2021 Hamburg

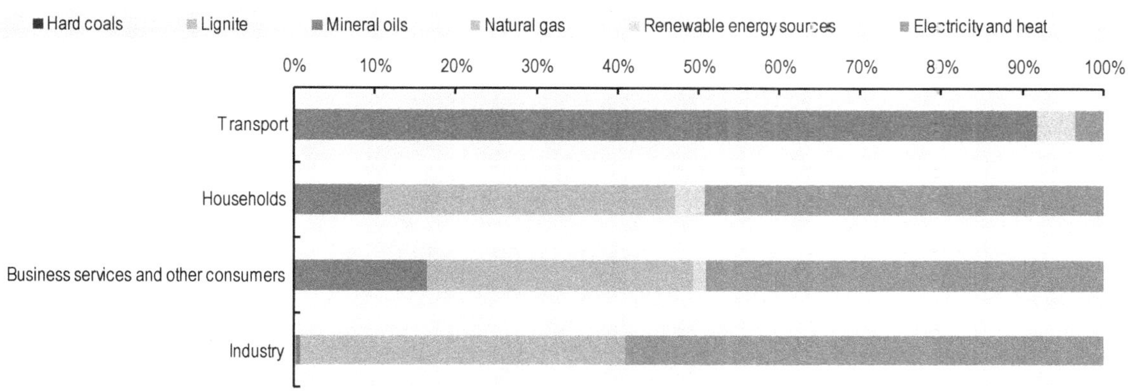

Source: (Hamburg Statistics Office, 2022[23]).

The metals industry, including steel, copper and aluminium production, is the largest energy user in the manufacturing sector, followed by food production (Figure 1.10.). Most energy use has not been electrified in the production of metals, which adds to transformation challenges. Natural gas is the largest source of energy for producing metal and metal products. It is used in basic iron and steel production. It needs to be phased out to reach climate neutrality by 2040. Electricity is dominant in most of the other subsectors (Figure 1.11).

Lignite and mineral oils are still used in a few activities, especially in motor vehicle production and the manufacturing of chemical products. Its use is marginal, contributing little to energy use and emissions. Oil refining is not included in Hamburg's industrial energy use statistics. However, it is the most energy-intensive manufacturing sector in manufacturing (OECD, 2023[7]). It also faces particular challenges because its output consists of fossil fuels which need to be phased out.

Figure 1.10. Final energy consumption of the manufacturing sector in 2021

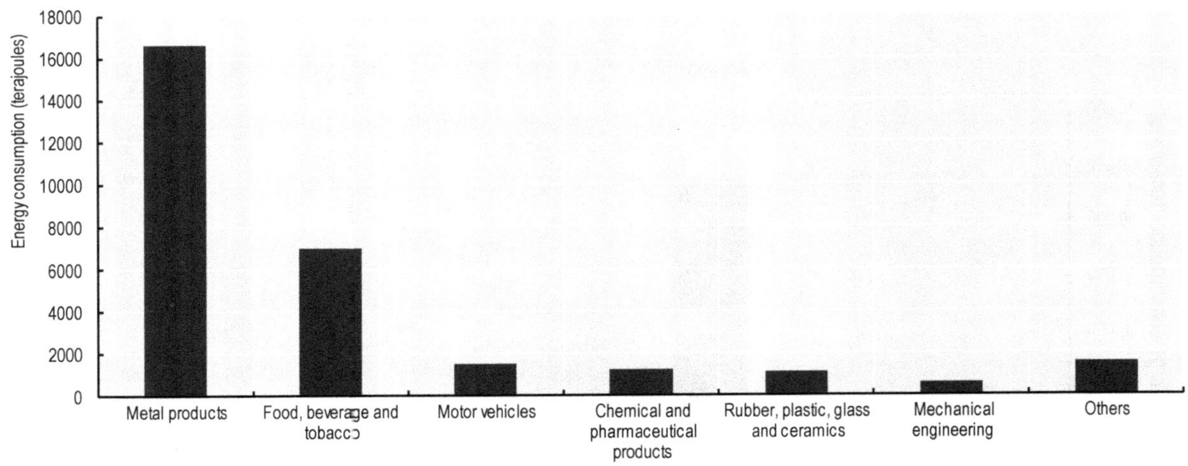

Source: Excludes oil refining. (Hamburg Statistics Office, 2022[23]).

Figure 1.11. Natural gas is a substantial source of emissions in metal production

Composition of final energy consumption in manufacturing by energy source, 2021

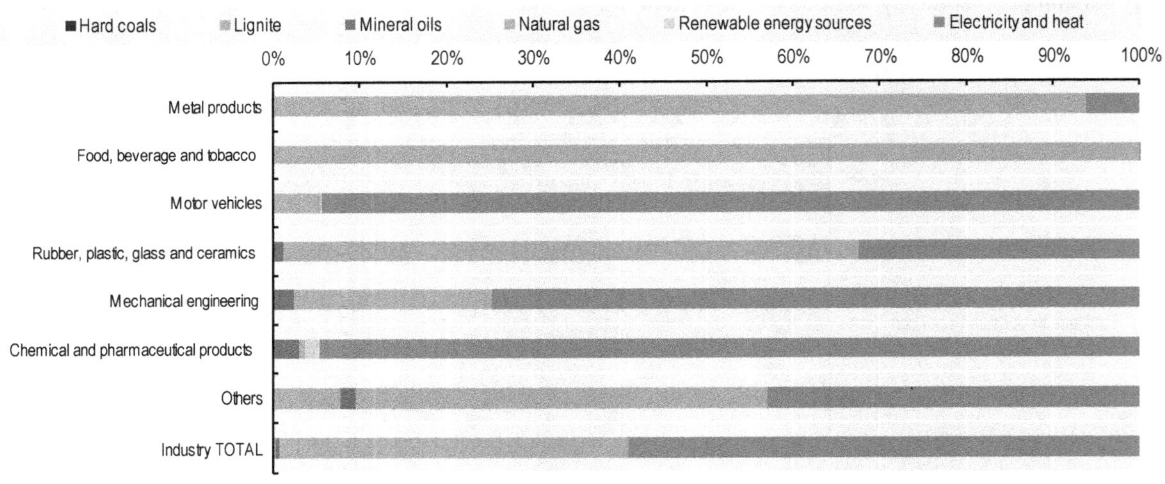

Source: (Hamburg Statistics Office, 2022[23])

The sectoral structure of the Hamburg economy

Hamburg is an economically successful, prosperous region by European and OECD standards. Its GDP per capita is the highest of the German regions, although growth has been weaker than in the most dynamic metropolitan areas in Germany as described in the OECD Hamburg Territorial Review (OECD, 2019[29]), Acting as soon as possible on the transition to climate neutrality can help position the Hamburg economy, making investment future-proof, avoiding unnecessary future costs, prepare for the challenges and benefit from the opportunities.

The Hamburg economy is broadly diversified with business services and manufacturing as major contributors to the generation of value-added (Figure 1.12). Transport and logistics, trade, and real estate sectors account for the largest service shares. Setting aside services dominated by the public sector, information and communication services and scientific and technical activities are also the most notable for their contribution to economic growth in 2021, as well as the manufacturing sector (Figure 1.13). By contrast, wholesale and retail trade, transport and logistics as well as financial services have shrunk or grown little. Manufacturing is also particularly energy and emission-intensive, as the industry share in energy consumption and Scope 1 and 2 emissions (mostly accounted for by manufacturing) exceeds the manufacturing share in value-added.

Figure 1.12. Business services and manufacturing contribute most to private sector value-added

Shares of economy-wide value added by sector in 2021 Hamburg (basic prices)

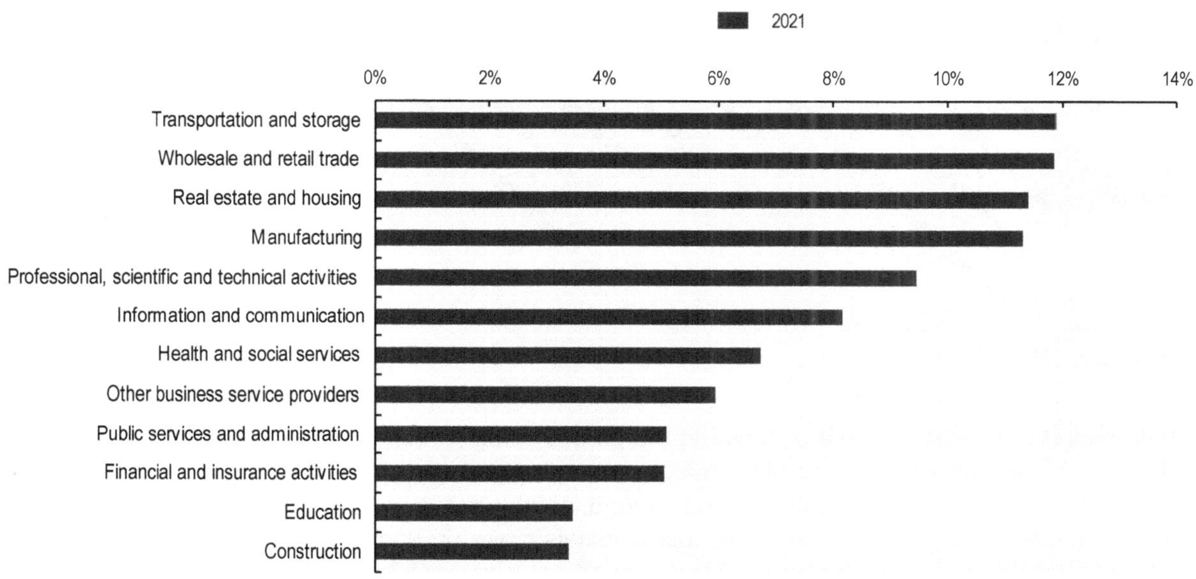

Source: Statistics Office Hamburg (2023).

Figure 1.13. Information and communication services, industry and business services have been growth drivers

Real value-added growth, Hamburg, 2015-2021

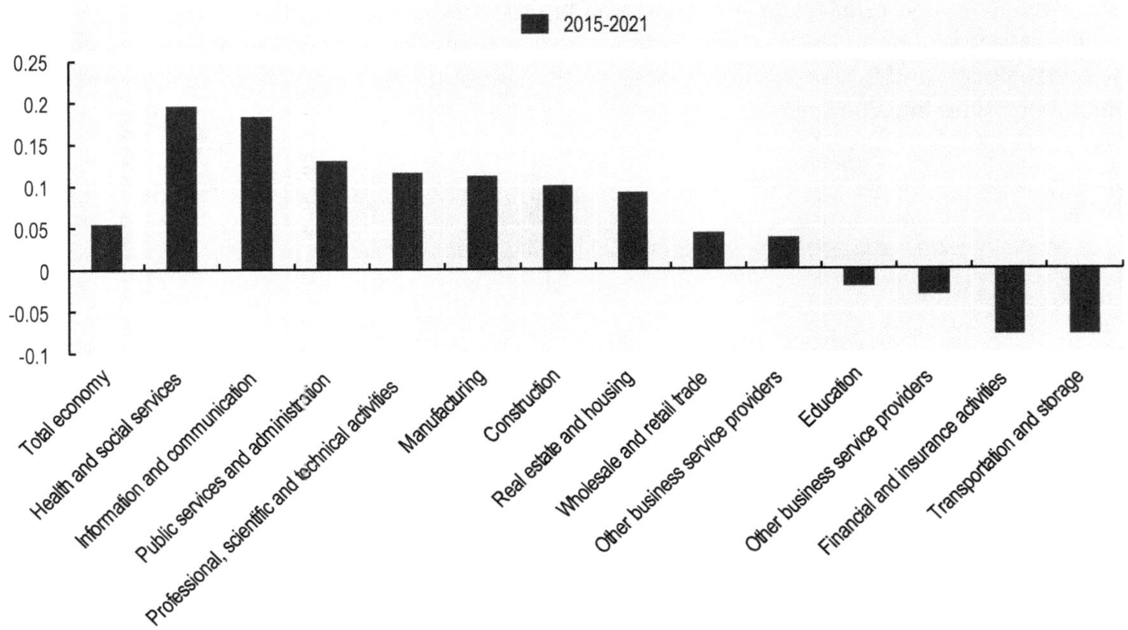

Source: Statistics Office Hamburg (2023)

While transport and logistics have declined in the past, the region continues to strongly rely on logistics and trade, with the Hamburg port one of the major economic pillars (Box 1.5). The port is closely intertwined with manufacturing industries in Hamburg and beyond, notably those producing basic materials (including aluminium, copper and steel). For example, basic metals make up about a quarter of the general cargo import and export business of the port, while transport vehicles contribute 40%. Manufacturing and port activity irrigate other sectors. This can include activity through both links in supply chains, such as trading, technical or financial services, as well as demand effects from the income they generate. This is reinforced by high labour productivity in manufacturing.

As discussed in chapter 3, the city's economy is well-placed to play a leading role in the decarbonisation of freight transport, which is difficult to decarbonise on a global scale, building on its strong rail infrastructure and its status as one of the 4 biggest ports in Europe. Moreover, shipping and rail freight are among the least energy-intensive freight modes. As this chapter has shown, transport is energy-intensive and cutting back on energy use is a key challenge in the transition to climate neutrality. Successful decarbonisation could therefore further add economic dynamism to the city. It is also well-placed in the decarbonisation of heavy-duty road transport as well as in the decarbonisation of key manufacturing industries. As discussed in Chapter 3, it can further play a major role as a hydrogen hub. It can draw on a broad range of forward-looking research and development and infrastructure deployment projects in these activities.

Box 1.5. Hamburg port and related transport services

Hamburg has the third biggest port in Europe, after Rotterdam and Antwerp. It is closely related to land-based transport from and to the Hamburg hinterland, notably Germany, central-eastern Europe and the Baltic region (Figure 1.14). 70% of Hamburg businesses in transport and logistics work in the land-based transport of goods. Container shipment carries a wide diversity of manufactured products and dominates port activity but has been declining of cargo shipments. China is by far the most important origin and destination country. General cargo provides higher value-added activity to the port and has been growing and is more closely linked to local manufacturing than container shipment. About a quarter of goods shipped in Hamburg are reloaded to or from further maritime transport. Close to half of land-based transport is handled by rail, the remainder by road freight, with land-based trade connections mostly serving EU countries in central and eastern Europe, including over long distances (Figure 1.14). Rail prevails for the more distant locations and the port development plan foresees steps to further expand the rail share. This is analysed in more detail in Chapter 3.

Figure 1.14. Container train connections with Hamburg's port

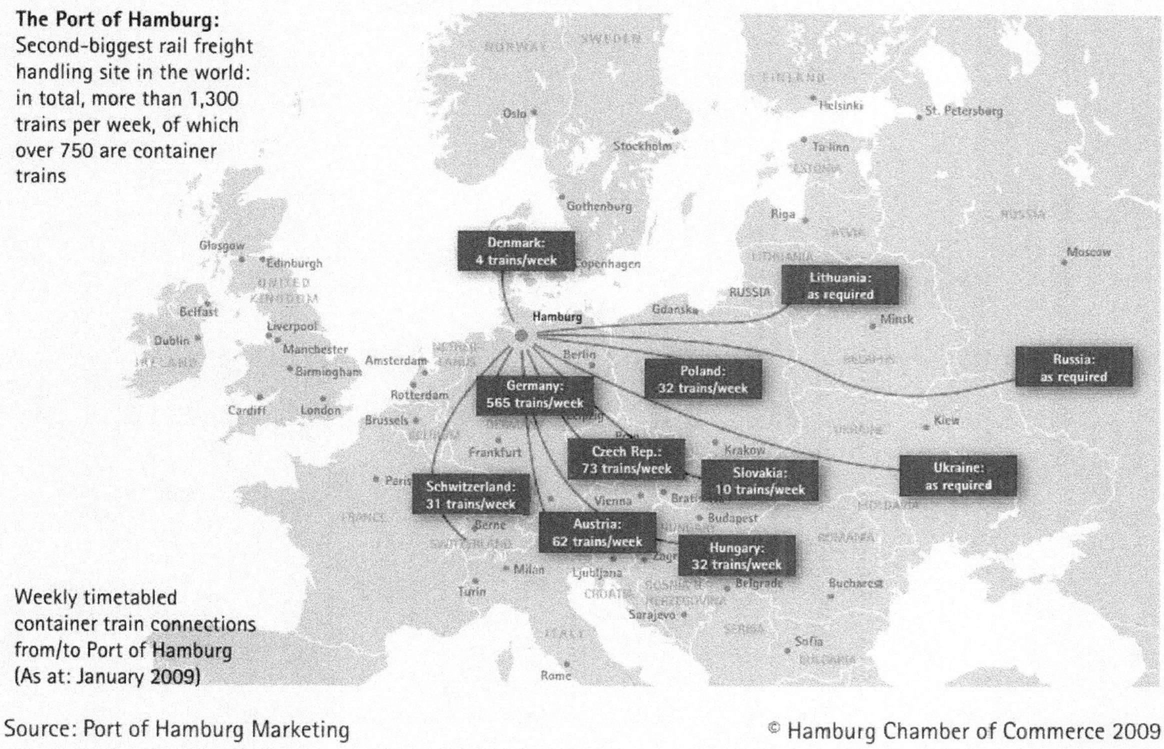

Source: (Hamburg Chamber of Commerce, 2010[30])

In many other service sectors, heating and cooling of buildings, electrifying equipment, and efforts to reduce energy use will dominate scope 1 emissions reductions. However, challenges to reduce Scope 3 emissions will differ. Some sectors process large volumes of emission-intensive goods and services. This applies to wholesale or retail trade or manufacturing. For example, Scope 3 emissions in the Hamburg-based copper production group *Aurubis* dwarf Scope 1 emissions (Chapter 2).

Construction activity will need to take on the task of making all buildings consistent with climate neutrality, which will bring a very large expansion of labour-intensive activity as well as new skill requirements. Business services as well as information and communication services will face opportunities, for example, from the use of digital technologies for the flexible use of intermittent renewables (Chapter 2) or the circular economy (Chapter 4).

OECD general equilibrium modelling suggests that sectoral value-added and employment shifts resulting from the transformation to reach climate neutrality are small in most sectors, across European Union countries, although there may be some loss of activity in some basic materials manufacturing industries. Impacts may be bigger in individual regions depending on their sectoral specialisation, but also their preparedness (OECD, 2021[15]).

Employment data allow a more detailed sectoral analysis than value-added data. Most sectors have expanded employment over the past 10 years. The wholesale and retail trade sector is the largest employer (Figure 1.15). Retail trade accounts for nearly 60% of the sector's employment (Figure 1.16). The contribution of manufacturing to employment is relatively small, reflecting its high productivity.

Figure 1.15. Wholesale and retail trade is the largest employer in the Hamburg economy

2011 and 2021 Hamburg Employment by Sector in Absolute Number (Thousands)

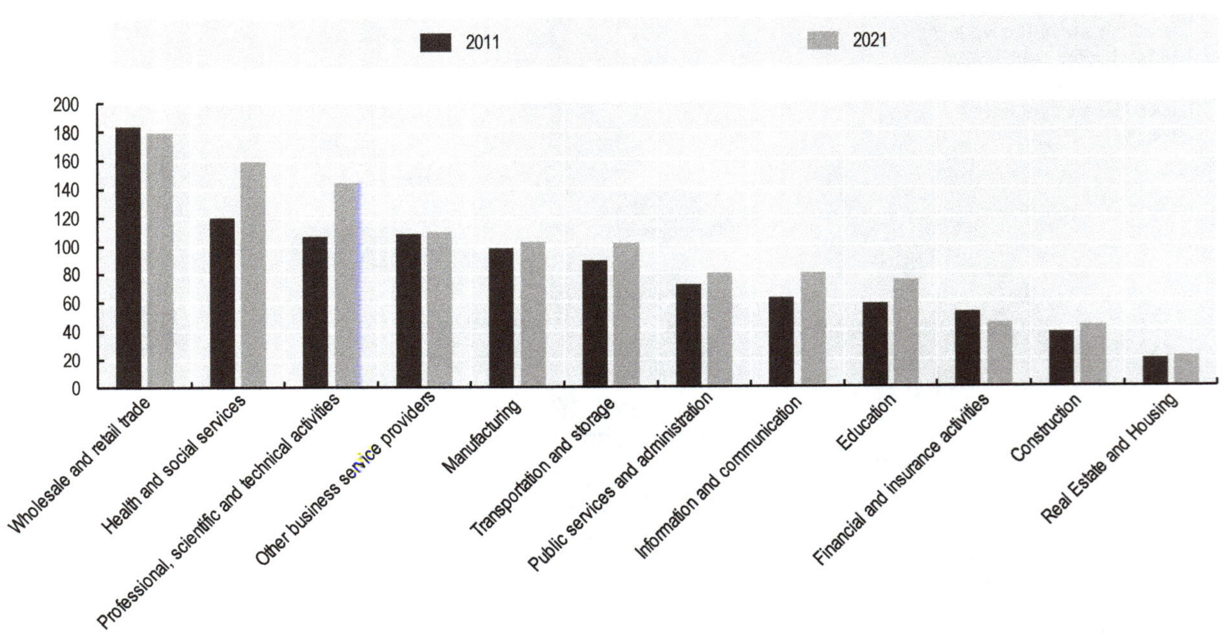

Note: Employment based on the domestic concept, counting workers according to the location of the workplace.
Source: Hamburg Statistics Office (2023).

Figure 1.16. Retail dominates employment in the trade sector

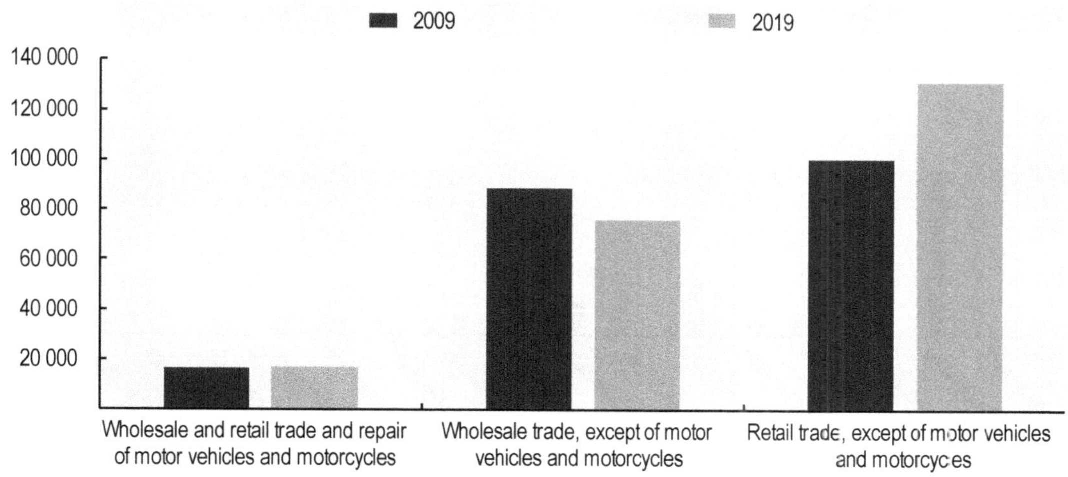

Note: Employment based on the domestic concept, counting workers according to the location of the workplace.
Source: Eurostat Structural Business Statistics (SBS) database (2022).

Transport equipment manufacture, repair and installation of machinery and equipment, and manufacture of machinery and equipment are the three most important employers in manufacturing (Figure 1.17.). Transport equipment mostly includes airplane production. Airplane production does not stand out in terms of local Scope 1 emissions and energy use, as shown above. It however faces major challenges to climate neutrality which result from emissions in flights. These are downstream Scope 3 emissions. Technologies for emission-free airplane fuels are not yet available. Beyond fuels, vapour trails and cloud formation of planes also contribute substantially to global warming. A near-term emission reduction option is the use of biofuels. In the longer-term substitution of air travel, especially short and medium-haul, may reduce demand.

Among the manufacturing sectors with relatively high Scope 1 emissions and high energy use, basic metals, oil refining and food production, each employ between 3 000 and 8 000 workers. However, only part of the food industry generates substantial Scope 1 emissions. These activities also face substantial challenges from Scope 3 emissions. These include the extraction of raw materials and downstream emissions from product use.

Figure 1.17. Airplane producers employ the most workers in the manufacturing sector

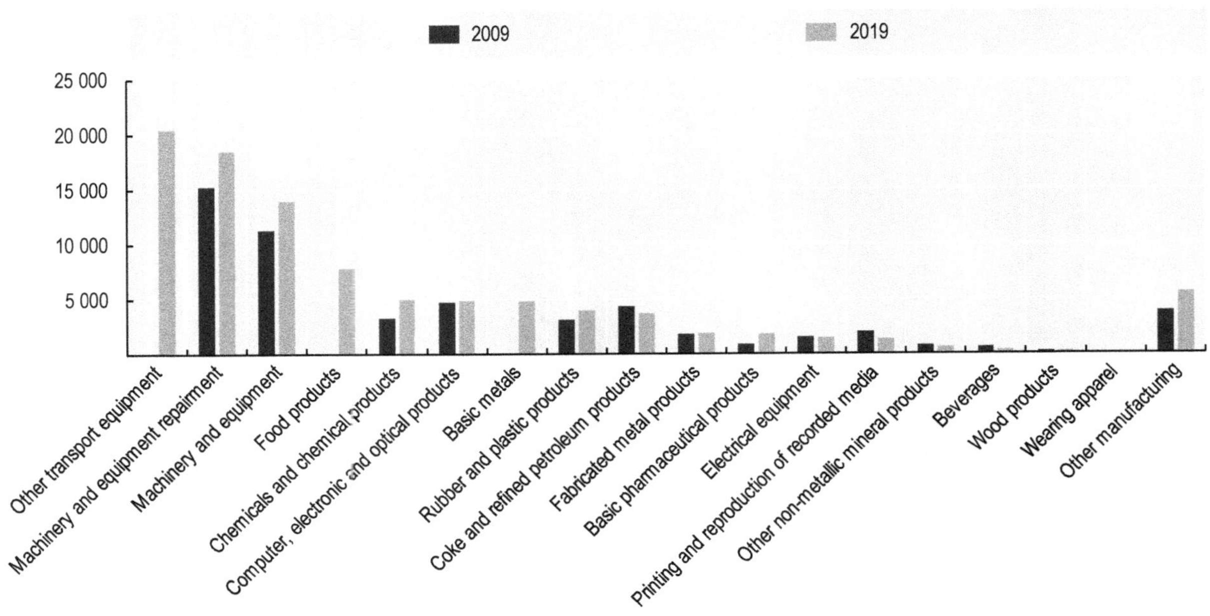

Note: Employment data for *Other transport equipment, Food products, Basic metals,* and *Wearing apparel* are not available in 2009. Airplane production accounts for most employment in *Other transport equipment*. Employment based on the domestic concept, counting workers according to the location of the workplace.
Source: Eurostat Structural Business Statistics (SBS) database (2022).

Warehousing and support activities account for about 50% of employment in transport and logistics (Figure 1.18). Businesses in support activities provide logistics services, services in loading and unloading of freight, support services for shipping and transport services across transport modes. These activities are strongly concerned by the transformations to climate neutrality and can benefit from ambitious action to make transport ready for climate neutrality (Chapter 3).

Figure 1.18. Warehousing and support activities dominate employment in transport services

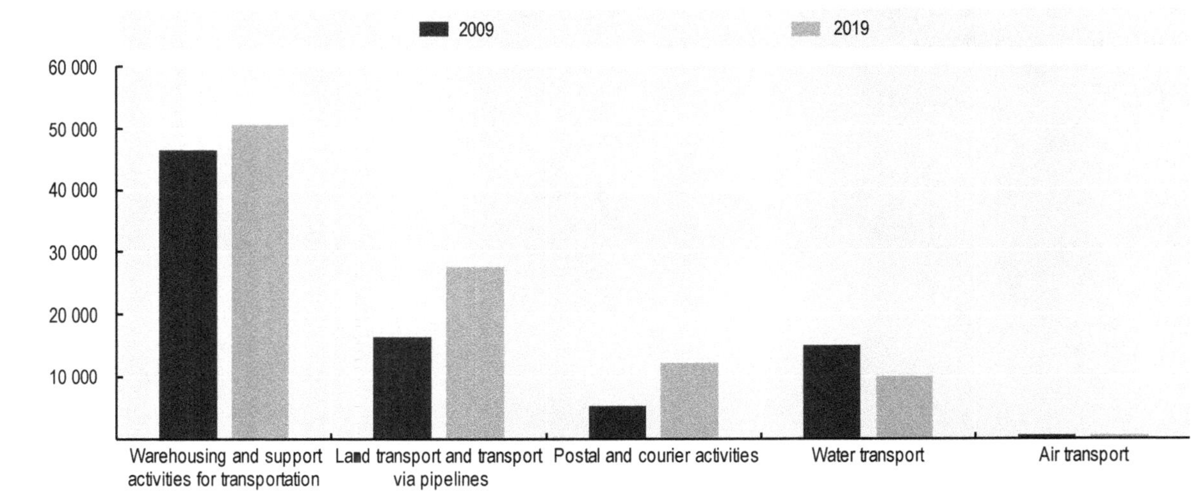

Note: Employment based on the domestic concept, counting workers according to the location of the workplace.
Source: Eurostat Structural Business Statistics (SBS) database (2022).

Climate neutrality action in comparison cities

This section draws lessons from climate action in comparison cities, from their targets and action plans for the decarbonisation of the Hamburg economy. The three selected cities are Rotterdam, Seattle, and Stockholm (Table 1.3). Each provides features to be analysed for the benefit of the Hamburg economy.

Table 1.3. Comparison cities table

	Rotterdam	Seattle	Stockholm
Climate Neutrality target date	2050	2050	2040
2030 Target	55% CO_2 emissions reduction compared to 2019	58% GHG emissions reduction compared to 2008	Fossil-free municipality operations by 2030 Reduce GHG emissions from transport by 70%
Port City	Yes, biggest in Europe	Yes	Yes, but mainly a passenger port
Population	624,00	733,919	923,516
Scope of emissions in Climate Action Plan	Scope 1 – port and industry, buildings, transport Scope 2 –energy Scope 3 – consumption of citizens	Scope 1 – transport, buildings, port Scope 2 –energy Scope 3 – the port	Scope 1 – buildings, transport Scope 2 –energy Scope 3 – consumption of citizens

Source: (City of Seattle, 2018[31]) (Energieswitch, 2019[32]) (Stockholms Stad, 2020[33]) (Port of Seattle, 2021[34])

Rotterdam has a similar economy to Hamburg, as Europe's biggest port and transport hub, hosting industrial activity, notably in oil refining. It therefore may face comparable challenges in the transition. Seattle also hosts an international port and provides a very detailed decarbonisation plan for it. It also hosts industrial activity, notably in airplane construction. Rotterdam and Seattle propose climate neutrality objectives for 2050, and Stockholm is more ambitious with a climate neutrality objective for 2040. All three climate action plans cover Scope 1 emissions of buildings and transport, Seattle and Rotterdam also focus on the emissions of the port and industry. All three cities also include Scope 2 emissions from energy use, targeting emission-free energy provision. Rotterdam and Stockholm also target Scope 3 emissions from the consumption of goods and services. Seattle focuses mainly on Scope 3 emissions from the port.

To understand the emission composition of the selected cities, the emissions estimates (Box 1.6) of the Functional Urban Areas (FUAs) are compared to those of the Hamburg FUA. The FUA is composed of the city and its surrounding local units that are part of the city's labour market (commuting zone) (Dijkstra, Poelman and Veneri, 2019[35]). FUAs are the most granular geographical breakdown available for which city GHG emission estimates are available.

Box 1.6. Estimating regional GHG emissions

Regional emissions are estimated based on the Emissions Database for Global Atmospheric Research (EDGAR) of the European Commission's Joint Research Centre (ECJRC). It allocates national Scope 1 GHG emissions from all sectors except emissions from land use, land use change and forestry, to locations according to about 300 proxies for 26 main sectors, further subdivided into subsectors, depending on the type of technology and International Energy Agency (IEA) fuel types, following IPCC reporting guidelines. Locations of emissions are identified with various sources of spatial research. The proxies capture a substantial part, but not all, of the local emission determinants.

Emissions are attributed as follows:

- **Manufacturing** includes emissions which are allocated to the plant location coordinates on point source grid maps. Government pollution and emission registers are the main sources for point locations. A specific proxy captures cement emissions for the world-leading producers of cement based on the plant locations and annual material and energy carrier flows.
- **Buildings** includes the energy use for buildings. Emissions are attributed spatially using high-resolution criteria on population and built-up density. The dataset classifies six categories of human settlements (mostly uninhabited rural, dispersed rural areas, villages, towns, suburbs, and urban centres) using satellite imagery. The data are combined with population density from updated population censuses. Emissions from fossil fuel combustion in the household and commercial sectors are attributed over total population density maps. The estimated spatial distribution of the emissions hence cannot consider subnational spatial differences in energy efficiency standards of building or fuel types.
- **Energy production** contains all combustion of fuels for electricity generation by power plants. Emissions are distributed according to the point source distribution data sets including intensity parameters, differentiated between fuel types (coal, gas and oil).
- **Energy extraction** includes process emissions and fugitive emissions during extraction and transport of fossil fuels. Gas flaring activities are distributed on night-time light data for areas with strong gas flaring activities, such as the North Sea region. The coordinates of coalmines help locate related emissions and distinguish between hard and brown coal.
- **Transport** encompasses freight and passenger ground, sea, and air transport. Transport route information is used for the spatial attribution of transport emissions. Proxy data for three road types worldwide (highways, primary and secondary, residential, and commercial roads) obtained from OpenStreetMap is combined with national weighting factors to distribute emissions for each road type. The distribution depends on the type of vehicles circulating on each road type, with data on traffic flows by road type to the extent available from regional sources, or imputation of traffic flow data based on population density. Similar data is used for railways and inland waterways. For maritime traffic, traffic identification and tracking data are used. For air traffic, data from the International Civil Aviation Organization, flight information, and flight patterns (landing/take-off cycle) are used and allocated according to the routes.
- **Agriculture** includes all agricultural and fishing activity, notably emissions from agricultural soils, agricultural waste burning, enteric fermentation, and manure management. Sources are attributed spatially according to agricultural land use, soil type, local livestock density, and crop type datasets and maps from the Food and Agriculture Organization (FAO). Fuel combustion emissions in the agricultural sector are distributed over "rural" areas (mostly uninhabited and dispersed rural areas) for all fuels, except for natural gas, which is assumed to be used mainly in villages. Emissions from fuel combustion in agricultural activity, while minor, are also estimated, for example, using maps of fishing activity.
- **Waste** includes emissions from waste incineration without energy recovery.
- **Other** contains NO_x and NH_3 emissions from nitrogen deposition, using geospatial information, cropland and grassland maps and arable land. It also contains emissions from fossil fuel fires which are estimated using data on oil production and coal fires.

Note: Population-based gap-filling techniques are used for residual emissions that cannot be located, especially in the industrial and power sector.
Source: EDGAR v6 (2018), (European Commission, 2022[36])

Rotterdam and Hamburg have comparable estimated Scope 1 emissions, in amount and sectoral contributions (Figure 1.19), with similar reduction challenges. Seattle has lower emissions on account of power and heat generation. Stockholm's estimated emissions are significantly lower, which may also be why their carbon neutrality goal is more ambitious. Stockholm has decarbonised power generation as well as heating in buildings, in large part owing to large-scale biomass firing. Biofuels also reduce emissions in road transport.

Figure 1.19. Scope 1 estimated emissions in the functional urban areas of Hamburg and Rotterdam are similar

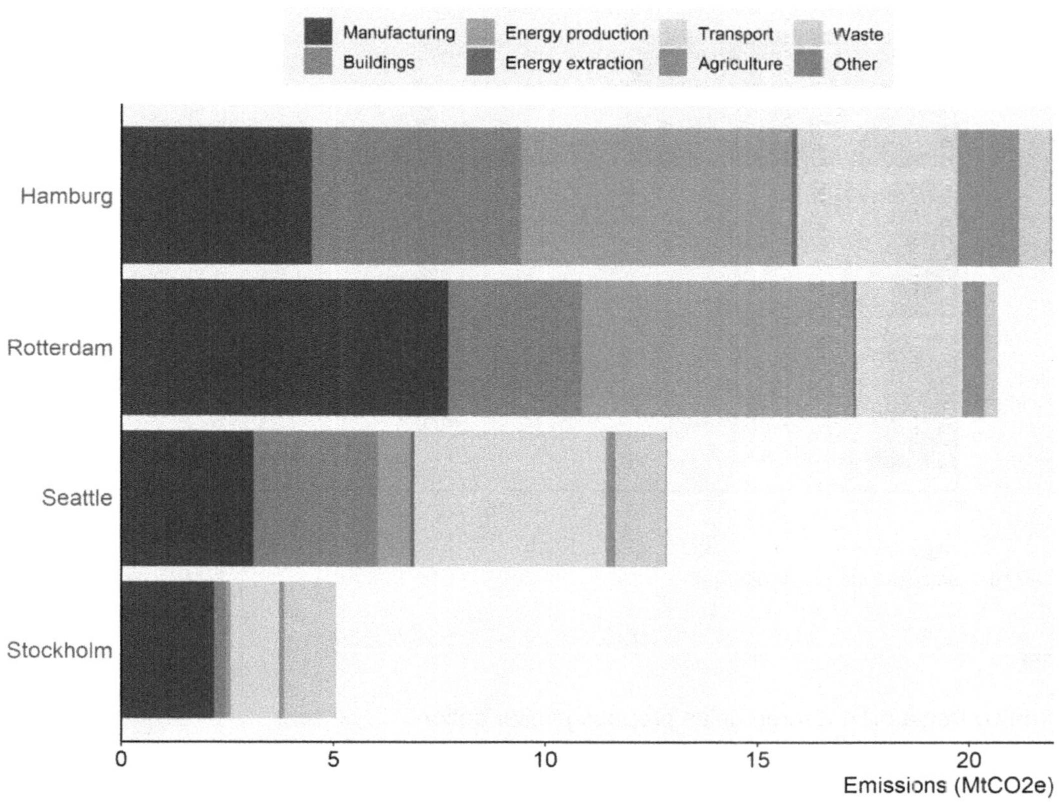

Source: EDGAR v6, (European Commission, 2022[36])

Hamburg's emissions per capita seem to be between Rotterdam's and Seattle's (Figure 1.20). Sectoral emissions per capita illustrate in which sectors bigger challenges remain. For example, the decarbonisation challenges of manufacturing in Rotterdam are bigger than in Hamburg. The transport sector of Hamburg has similar emissions per capita to those of Seattle, but they are lower in Stockholm and higher in Rotterdam. The emissions per capita arising from waste are similar across all four cities.

Figure 1.20. Estimated Scope 1 emissions per capita in the Rotterdam and Hamburg functional urban areas are higher than in Seattle or Stockholm, 2018

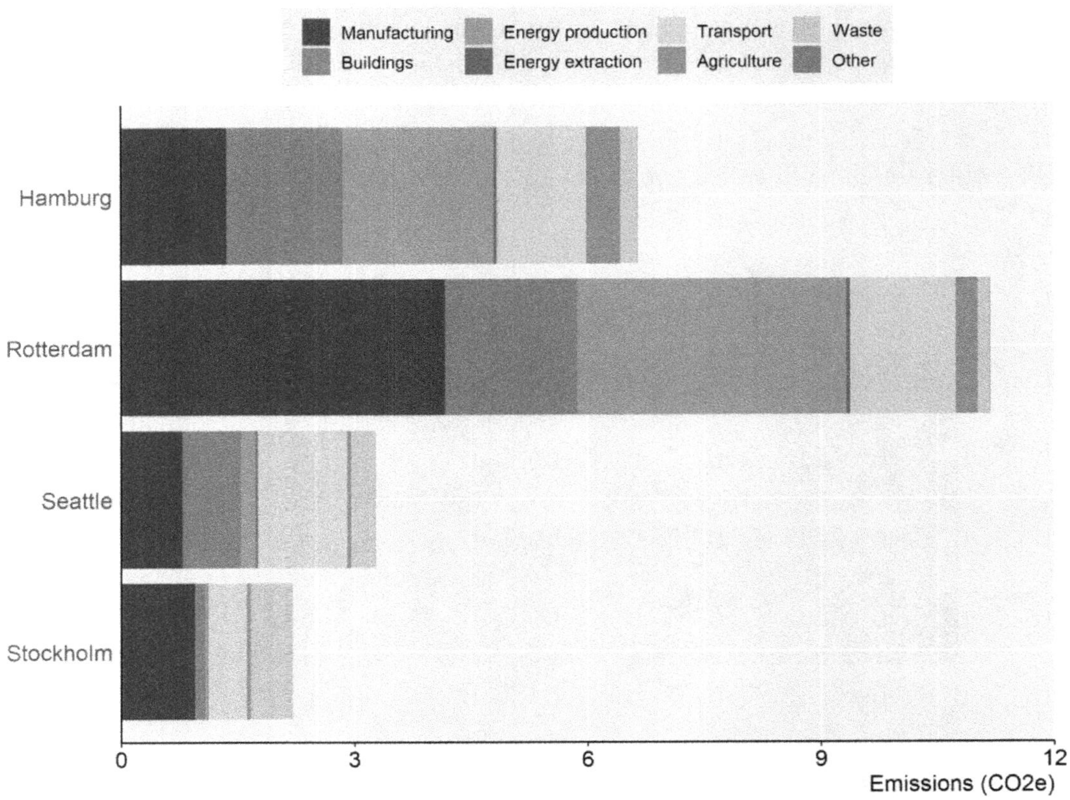

Source: EDGAR v6, (European Commission, 2022[36])

Transport

To decarbonise transport the three cities propose similar actions. They are:

- **Mobility hubs** - the main aim is to aggregate connections between different transport modes, to improve their interoperability, with a focus on public transport, shared and e-mobility. The idea is to aggregate transport connections, so that mobility options carpools, electric cars, shared bicycles, and public transport are all connected to each other. Expected benefits are better transport and mobility access in low-income neighbourhoods as well as a reduction in car use.
- **Electrification of transport** – which includes building charging points and rolling out shared electric mobility throughout the city. Seattle is mapping the optimal distribution of charging infrastructure to ensure equal distribution throughout the city. Stockholm has the goal of at least 4,000 public charging points. Stockholm also plans to introduce electric trucks for optimised inter-city deliveries. They will at the same time collect waste, thereby reducing traffic.

There are also actions which are particular to each city. For example, Rotterdam is placing a focus on creating a business leaders' roundtable, including businesses with 110,000 employees altogether, to discuss how to improve mobility and share best practices. The city approached businesses to reach targeted agreements about sustainable commuting and business traffic with a minimum of 50% CO_2 reduction by 2030.

Rotterdam and Stockholm are also focusing on finding emission-free solutions for the supply of building materials and making more efficient use of vehicles for delivery and collection. For Stockholm, this means developing an underground network and using sewage tunnels to transport building materials by boat instead of truck. To reduce the number of light trucks, Stockholm is also working with e-commerce distributors to create optimal routes to drop-off points, to reduce the number of trips. Rotterdam instead is consulting with a transport service provider to transition to emission-free inner-city heavy-duty transport. Optimised cargo logistics processes include decoupling points, where delivery trucks swap bodies moving containers to light vehicles or the use of plug-in hybrid trucks (City of Rotterdam, 2020[37]). Rotterdam is also working to reduce the number of transport movements of commercial waste vehicles.

Buildings

To reduce building emissions the cities focus on fossil-fuel-free buildings, energy efficiency and public participation.

- **Fossil-fuel-free buildings** - Rotterdam is working with neighbourhoods, property investors and housing associations for tailored plans to make all buildings natural gas free. Most neighbourhoods aim to be natural gas free by 2030. This includes connecting buildings to a district heating network as well as investing in the insulation of buildings, in climate adaptation and in opportunities for circularity. Seattle is also attempting to move buildings away from heating oil by supporting the conversion of oil-heated homes to electricity. Currently, there are recommendations to convert 18,000 homes from heating oil to electric heat pumps and help finance homes that are unable to switch on their own. Stockholm aims to fully phase out fossil oil and coal by connecting to heating plants with district heating and using biofuels. The city of Stockholm provides energy advice to property owners.
- **Energy efficiency in buildings** - Rotterdam has a lot of programs to optimise performance such as the installation of green roofs, measures against heat stress, to start pilot energy cooperatives, insulate homes, and make installations and lighting more sustainable. There are plans for 1000 homes to be equipped with solar panels and expand electric cooking. Seattle aims to provide property owners with advice on energy efficiency. The Seattle City Lights program aims at energy savings through Energy Efficiency as a Service (EEaS). EEaS helps overcome split incentive barriers in commercial buildings, where there is little motivation for owners or investors to finance retrofits which benefit the tenants. Tenants pay for the provision of energy-saving investments. EEaS lets investors finance projects with predictable returns, owners generate a new revenue stream, and tenants occupy energy-efficient spaces.
- **Public participation** - Rotterdam has set up a Climate Roundtable for the Built Environment and is setting up a digital platform to exchange knowledge and advice on the decarbonisation of buildings. Rotterdam is also designing a toolbox for real estate agents to educate and inform customers about opportunities for going natural gas-free and promoting sustainability. The city of Seattle intends to work with building owners through incentives and technical assistance to help them become voluntary early adopters and phase in performance requirements.

Consumption-based emissions

All three cities have some focus on consumption emissions, with specific focuses on food and waste. Stockholm also mentions actions to tackle aviation emissions.

- **Food** - All three cities are looking for opportunities to reduce emissions from food consumption, mainly through the reduction of food waste. Rotterdam has made agreements with producers and other parties to avoid waste by providing meals to social restaurants. Rotterdam is also piloting a study to understand how the city can shift consumer preferences towards plant-based nutrition.
- **Circular economy and waste prevention** – Rotterdam is focusing on the reduction of textile waste, hence has opened a clothes exchange, where 2nd hand clothes are traded. Rotterdam is

also looking into developing a circular department store, allowing consumers to find sustainable brands in one place. Rotterdam is also researching opportunities for more and better recycling of textiles and developing a chemical recycling facility for local upgrading of discarded textiles. Seattle is focusing on building deconstruction while saving building materials. Stockholm is developing a digital system to make recycling operations more accessible to the public.

- **Aviation** – Stockholm is running a city-wide communication campaign on the impact and alternatives of air travel. The city is also researching the most effective carbon offsets to mitigate the emissions from air travel.

Clean Energy

The cities of Rotterdam and Stockholm also have separate climate actions to accelerate the uptake of clean energy (wind and sun) in the city and reduce reliance on fossil fuels.

- **Wind** – The city of Rotterdam is working on accelerating four-wind energy projects and there is a consultation ongoing for the North Sea Program (2022-2026) with the goal of an additional 10GW of wind at sea by 2030. The important part is to connect wind electricity to the national platform. It could serve to meet regionally concentrated electricity demand for industrial purposes, including hydrogen production (see below). Stockholm plans on buying electricity from wind turbines through long-term contracts.
- **Solar** – The city of Rotterdam has a pilot project to install photovoltaic panels on residential roofs through energy cooperatives, with the goal of 90 solar roofs by 2025. This includes knowledge sharing, networking, development and management of projects. For residential areas, the city plans to research and develop links between the generation, storage, and distribution of solar electricity. The project also aims to put solar panels on all suitable company and parking roofs. A subsidy scheme started in mid-2021 to enable entrepreneurs to benefit from a roof capacity assessment for solar panels. In Stockholm, estimates indicate that potential electricity produced with photovoltaic panels could cover more than 10% of the city's needs. The aim is to increase solar production by 100% relative to 2018.
- **CCS** – The city of Stockholm aims to introduce bioenergy use combined with carbon capture and storage (BECCS) to reach net negative emissions. In Stockholm, the combined heat and power plant used for district heating would be suitable for CCS. A trial is planned to be built at the plant.

The port and industry

Rotterdam

The Port of Rotterdam is among the top 15 worldwide in terms of cargo throughput and containers shipped. The economic and cultural roots of the city are closely connected to the port (OECD, 2016[38]). In 2018. the port of Rotterdam contributed 6.2% of value added to the economy of the Netherlands (Port of Rotterdam, 2018[39]). The port and industry cluster provide direct and indirect employment for almost 400,000 people (Energieswitch, 2019[32]). The goal of the city of Rotterdam is to become the most sustainable port in the world. The Climate Roundtable of Port and Industry includes the authority of the Port of Rotterdam, energy companies, the municipality and provinces of Zuid-Holland, the nature and environmental federation of Zuid-Holland, companies that form the Port of Rotterdam industrial complex, and other government and knowledge institutions.

The Climate Roundtable of Port and Industry established phases and objectives in the decarbonisation of the port and industry. They concern efficiency measures, transitioning from fossil fuels to sustainable energy sources and creating economic and employment opportunities, putting Rotterdam in a strong competitive position.

The Climate Roundtable of Port and Industry have developed an investment agenda, which includes two main projects:

- **The Cluster Energy System (CES)** identifies key infrastructure necessary for the transition, including hydrogen infrastructure, wind farms, and CCS infrastructure with multiple pipelines (Box 1.7).
- **The Data Safehouse** is an exchange of information between major industrial companies and electricity distribution network operators to prepare for the electrification of industrial energy needs. The aim is to allow network operators to understand how much extra electricity is required to plan investment efficiently and meet the needs of companies in the switch to renewables.

The projects are also linked to an **acceleration platform** which aims to provide support in finding funding and removing barriers in legislation and regulations. The platform aims to accelerate the implementation of hydrogen, industrial electrification, industrial residual heating and circular processes projects.

Box 1.7. The six projects identified by the Cluster Energy Strategy (CES) in Rotterdam

- The **pipeline corridor** (*Delta corridor*), which will host pipelines for hydrogen, CO_2, liquefied petroleum gases (LPG) and possibly circular raw materials. The project is focused on strengthening the infrastructure of Rotterdam via Moerdijk/Geertruidenberg to Geleen and the connection to North Rhine-Westphalia.
- **Infrastructure for transport and storage of CO_2** *(The Porthos projects)* in depleted gas fields under the North Sea. This would connect the port via pipeline to a compressor station, and then into an empty offshore gas field. Once operational, it would store 2.5m tonnes of CO_2 annually, for 15 years, equivalent to nearly 2% of Dutch emissions.
- **Construction of a hydrogen main backbone pipeline** (*HyTransPort.RTM pipeline*) through the port area. Local production, imports of huge hydrogen volumes and transit to the hinterland will be integrated. This project was developed by *Gasunie* and the Port of Rotterdam Authority.
- **Infrastructure for low-carbon hydrogen production and transport** *(H-vision)* to be used as fuel in industry. Three product pipelines are planned for the supply of hydrogen, for the discharge of CO_2, and for the transport of low-carbon hydrogen to industrial consumers. H-vision also aims for two hydrogen production plants (plant 1 in 2027 and plant 2 in 2032). The project aims to reduce emissions by 2.7 million tonnes annually by 2032. 90% of low-carbon hydrogen will be produced with residual methane for example from refineries, supplemented by a small share of natural gas (H-vision, 2017[40]).
- **Pipeline for the transport of heat** *(Warmtelin Q)* generated by the port and industry to Rotterdam to homes, offices, and greenhouse horticulture in the region.
- **Extra landing power from offshore wind farms** combined with plans by grid operators to create higher capacity. This should bring the needed green power for green hydrogen production and electrification in industry.

Source: (Port of Rotterdam, 2021[41])

The port also aims to produce renewable energy. According to The Climate Roundtable of Port and Industry, the additional sun potential is 130-150 MWp. The port has committed to installing solar panels on commercial roofs within the port area. The port also aims to build a floating solar park but this is currently postponed due to financial barriers.

The industrial processes in Rotterdam are to be electrified to replace natural gas. The Field Lab Industrial Electrification allows the industry to gain knowledge about potential electrification, test new technologies and make it ready for implementation. Furthermore, the heating from residual heat in industry, heat from geothermal energy, and other local heat sources are expected to provide heat to businesses in the region, Zuid-Holland.

The port of Rotterdam also has circular economy plans. The port aims to utilise residual flows, biomass, and captured CO_2 in industrial processes. The industry in the port of Rotterdam will eventually be based on circular and renewable carbon materials, notably from biomass sustainable biomass and hydrogen. New factories and value chains organised in clusters will support the raw materials transition.

Labour market shortages are one of the biggest challenges faced by the port and industry, notably a shortage of technical personnel. Rotterdam aims to implement a training agenda to increase the required skills supply for the energy transition, coordinated by the Rotterdam Apprenticeship Agreement.

Seattle

The port of Seattle has its own climate targets and plan. The port aims to:

- **By 2030**, reduce Scope 1 and 2 emissions by 50% below 2005 and **by 2050** to be carbon neutral or carbon negative.
- **By 2030**, reduce Scope 3 emissions by 50% below 2007 and **by 2050** to 80% below 2007 levels.

The action scenario identifies strategies to reduce emissions that are directly and indirectly controlled by the port.

Reduction of Scope 1 emissions

To achieve immediate emission reduction in the boat fleet of the port, the initial plan is to switch to nonpetroleum-based fuels such as waste cooking oil and grease or other renewable feedstock. The current focus of the port is renewable diesel since it is more readily available than renewable gasoline. The goal is to shift to electric vehicles focusing first on light-duty vehicles while tracking developments in heavy-duty electric vehicles. The port will prepare for this transition by installing the required charging stations. At the same time, the port is eliminating under-utilised vehicles from the fleet and maximising use per vehicle.

To reduce waste the port aims to maximise the separation of common recyclables and organics, minimise solid waste generation and expand specialised recycling. The port will run waste audits, every three years, to assess proper waste disposal and develop site-specific reduction plants. The aim is also to identify items that are potentially recyclable but that are not accepted by the City's recycling program, such as scrap metals, building materials, electronics and furniture and add customised recycled programs when feasible.

Reduction of Scope 2 and 3 emissions

The port recognises that it has influence but no direct control of the emissions in maritime transport. This includes emissions from ships, harbour vessels, trains and other equipment, which account for 94% of the port's Scope 2 and 3 emissions. The port is envisioning what the sector will look like in a carbon-neutral economy and preparing the necessary infrastructure to be ready when the transition occurs. The port is encouraging the stakeholders, the community, industry and government, to shift towards the carbon-neutral vision. The goal is to provide these stakeholders with guidance and influence decisions through partnerships, programs, and port facility lease terms. The port is willing to play a leadership role by advocating for new technologies and fuels by supporting pilot projects and adopting small-scale zero-emission technologies in Port-owned workboats and cargo-handling equipment. The Hamburg port also has developed a vision for climate neutrality, setting a target for reaching it in port operations by 2040, with a planned monitoring of emission reductions towards the target.

The port also encourages start-ups in port-related industries to partner with the port's maritime innovation centre to achieve emission reduction in the maritime sector. It will also support workforce development and training to operate and maintain zero-emission maritime equipment.

The port is aiming to provide the infrastructure necessary for zero-emission vessels by 2030. This infrastructure includes new capacity for emission-free port manoeuvre boats, charging infrastructure, fuelling needs, and infrastructure for zero-emissions trucks. Until zero-emission vessels are developed there needs to be continuous improvement in vessel efficiency. The efficiency gains may occur through improved ship design and operational practices. The port will also support the adoption of zero-emissions cargo handling equipment by 2050, which involves replacing diesel-powered units. The port will also coordinate with cruise lines to evaluate carbon offset programs.

Other Scope 3 emissions the port is tackling are emissions commutes from port workers. 53% of port employees commute individually by car. To reduce the emissions generated, the port will encourage flexible work arrangements to reduce commuting days. It will also promote alternative modes of transport through subsidised vanpooling, bike sharing or organised carpooling.

Summing up: Lessons from comparison cities

Insights from the comparison cities yield recommendations for coordinating and facilitating climate actions in Hamburg. They are:

- **Business and public participation in decision-making.** This can come in the form of roundtables for specific sectors. Experts in the sectors and topics are also part of the decision-making process. It may also take the form of online platforms, information campaigns and welcoming feedback from citizens. The involvement of citizens, business owners and local stakeholders lowers resistance to the transition. Such participation needs to be organised as quickly as possible, as it may lengthen the time to prepare decisions.
- **Long-term vision** is essential. This ensures that long-term investments that businesses will undertake are in line with the climate neutrality targets. Rotterdam has done this through the development of an investment agenda for the port and industry Seattle demonstrates their long-term vision with the emission-free infrastructure investments of the port.
- **Continuous tracking and evaluation of actions**. Seattle is running multiple audits for the different climate actions the cities and ports are undertaking. This allows for a continuous reassessment and improvement of the ambition and climate action.
- **Reduction of Scope 2 and 3 emissions** by major actors.
- **Avoid increasing inequalities** within the city. Seattle is mapping the optimal distribution of electrical charging points to ensure that lower-income neighbourhoods are not left behind. Rotterdam is providing a subsidy scheme for the installation of solar panels which allows most citizens to participate.

Making the most of co-benefits from climate action

Local climate policies not only contribute to achieving global, national and regional climate goals but can also enhance the well-being of residents, workers and firms, substantially, especially in cities, in a broad range of dimensions (Box 1.8). From the local business perspective, well-being gains, such as from reduced traffic congestion or cleaner air, influence worker and firm location decisions and therefore have the potential to make Hamburg more attractive and competitive to businesses and workers. Economic outcomes also improve with health – for example lower air pollution boosts productivity (Dechezleprêtre, Rivers and Stadler, 2019[42]). Air pollution reduces performance in tasks requiring high skill levels, such as the performance of investors on the New York Stock Exchange (Heyes, Neidell and Saberian, 2016[43]).

Good accessibility of jobs and key facilities from homes, for example with on-demand mobility services, save travel time and make commutes less exhausting. A more convenient neighbourhood with such features would further attract skilled workers. Local climate policy can deliver these co-benefits.

These well-being benefits typically exceed the cost of local climate action. Several studies find that air quality co-benefits alone offset a large proportion of climate policy costs (Karlsson, Alfredsson and Westling, 2020[44]). For the East-Asia region, the co-benefits of climate change mitigation in terms of human health have been estimated to reach 6% of GDP, when also including the impact on climate adaptation. This exceeds the estimated cost of mitigation of 2% of GDP (Xie et al., 2018[45]). Many of the well-being benefits accrue locally and, unlike the climate benefits, immediately. They can therefore substantially improve the political economy of climate action, escaping the prisoner's dilemma perspective, and be a powerful motivator for local and regional action, including by the business community.

Passenger transport is strongly related to congestion, air and noise pollution, and car accidents, especially in urban areas. The social cost of private car use in cities is estimated to be about 6-7 times higher than the cost borne by individual car owners and drivers. (van Dender, 2019[46]). The high share of external costs due to congestion further highlights the major benefits from reducing the volume of car use (Table 1.4). Compared to conventional vehicles, lightweight electric vehicles emit only 18-19% less PM2.5 from non-exhaust sources (OECD, 2020[47]), which means electrification alone cannot eliminate air pollution, in part reflecting pollution from wheel friction. Thus, decarbonisation of passenger transport by reducing individual car use has a high potential for creating co-benefits. It can be done making public transport more attractive; and by designing pedestrian- and cycle-friendly public spaces. As argued below, ride-sharing can also make a major contribution. Ride-sharing goes well beyond car-sharing: it refers to the use of cars by several passengers with different trip origins and destinations. As argued below, ride-sharing should replace individual car use, resulting in the abolition of day-to-day individual car use, as discussed further below.

Table 1.4. Marginal external cost estimates of car use

Low and high estimates of marginal external cost of car use in Europe (prices of 2010, gasoline car)

	Euro-cent per vehicle-kilometre		%	
	Low	High	Low	High
Climate change	1.5	3.3	4.8	11.6
Pollution	0.4	1.1	1.5	3.9
Congestion	18.5	18.5	71.2	64.9
Accidents	4.8	4.8	18.5	16.8
Wear & tear	0.8	0.8	3.1	2.8
Noise	0.0	0.5		
TOTAL	26	28.5	100.0	100.0

Source: (van Dender, 2019[46])

Further benefits from reducing car use come from lower energy use, especially in the context of the likely more regional determination of electricity prices in the future, and lower indirect emissions. The transport sector, including transport and freight within Hamburg, is the largest energy consumer in Hamburg (Figure 1.7 above), and car use is likely to represent the largest share. Electric vehicles (EVs) are increasing demand for green electricity. Life-cycle emissions of EVs are also important. Therefore, decreasing the number of cars on the road is a robust approach to fundamentally decarbonise the transport sector. Lower energy demand is a key priority on the way to net-zero emissions.

Box 1.8. Co-benefits of local climate policies

Reducing air pollution

Inhalable small particulate matter (PM2.5) causes about 422 000 premature deaths in OECD countries annually and welfare loss of around 3% of GDP. Major disease effects include cardiovascular and respiratory diseases, such as stroke and COVID-19. It affects children's health the most (WHO, 2018) and also causes old-age dementia (Bishop, Ketcham and Kuminoff, 2018[48]). Education outcomes for children exposed to higher air pollution are substantially and lastingly lower (Heissel, Persico and Simon, 2019[49]). Moreover, air pollution reduces worker productivity, reflecting illness, but perhaps also cognitive performance (Dechezleprêtre, Rivers and Stadler, 2019[42]). Increasing active mobility reduces diseases and emissions, with an approximately 11:1 benefit-cost ratio (Chapman et al., 2018[50]).

Reducing noise pollution

Chronic exposure to noise levels above the WHO standard causes 12 000 premature deaths per year worldwide. 6.5 million people suffer from chronic sleep disturbance and 22 million people from prolonged high levels of annoyance due to noise pollution from transport or industry (EEA, 2020[51]). Noise exposure also affects patient outcomes and staff performance in hospitals as well as impairs cognitive performance in schoolchildren (Basner et al., 2014[52]). Electrifying vehicles and encouraging walking and cycling would significantly decrease noise pollution.

Reducing congestion

The cost of traffic congestion includes time loss as well as productivity losses from higher costs in the exchange of goods and services, especially within highly productive functional urban areas. Congestion hinders the region's socioeconomic development and raises the cost of doing business. Costs in high-income economies are estimated at 1% in Europe and between 0.7% and 0.9% in the USA. Cities in middle-income countries are substantially more congested due to less-developed public transport.

Healthier life from active mobility, road safety and better insulated buildings

It is estimated that if all Londoners walked or cycled for 20 minutes a day, the savings from public health spending could be up to approximately 496.4 million British Pounds. On average 11.5 kilometres of provisional pop-up bike lanes have been built per city in 106 European cities and each kilometre may have increased cycling by 0.6%. Every kilometre of cycle land produces annual health benefits of about USD 2 million, so investment may often pay off in less than a year. Active mobility policies also increase road safety. Health benefits of building energy efficiency investment subsidies in New Zealand have been estimated to pay off the costs (Grimes et al., 2012[53]). In the presence of energy poverty, the health benefits multiply the costs of loft insulation and are almost equal to the cost of wall insulation (Frontier Economics, 2017[54]).

Source: (OECD, 2021[15])

Harnessing co-benefits from reduced individual car use with ride-sharing

Expanding public transport is a good way to reduce car dependency. Access of residents in the Hamburg metropolitan area to other residents in the area is less good than in some other cities (Figure 1.21). This may not reflect differences in public transport service quality, but, instead the extent to which low-density rural areas are part of metropolitan travel to work areas. The Hamburg metropolitan region includes such low-density areas extensively. Low-density areas are typically less well served by public transport. Indeed, adding fixed-route buses and metro services at large enough volumes can be costly, especially in areas of lower population density, such as in the suburbs. Thus, harnessing the co-benefits from reduced individual car use requires a comprehensive policy package including housing policy, urban planning and a transport system that increases accessibility.

Figure 1.21. Public transport performance is weaker than in other metropolitan areas

Public transport performance of metropolitan areas, 2018

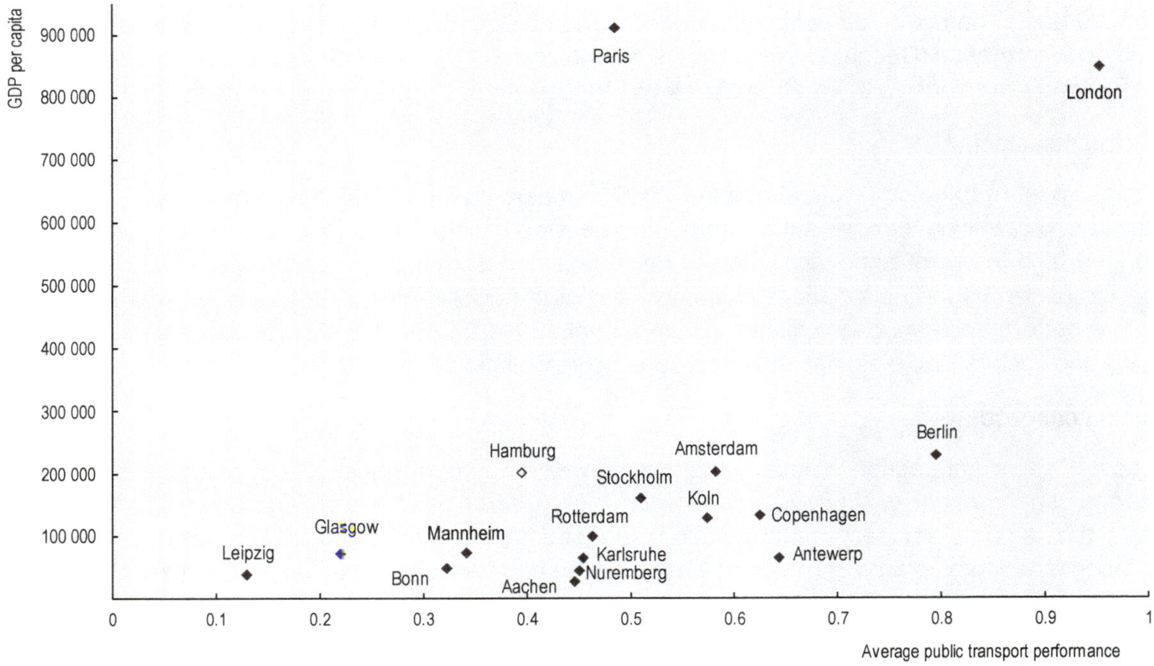

Note: Public transport performance is an index which reaches one when all residents within a radius of 8 km can be reached within 30 minutes from any point within the Hamburg metropolitan area. The metropolitan area is defined as the area where most people commute to the city of Hamburg. For GDP per capita (PPP) is indicated in USD per head, with constant prices from 2015. more detailed explanation, please refer to the 2019 ITF report 'Benchmarking Accessibility in Cities – Measuring the Impact of Proximity and Transport Performance'.
Source: OECD statistics and ITF.

StatLink https://stat.link/w3ir9y

Metropolitan governance will allow the residents to benefit from public transport and housing that are coordinated throughout municipalities in the same travel-to work area, while improving accessibility of jobs and services, reducing air pollution and congestion as well as eliminating GHG emissions. A denser and more contiguous residential development would help reduce emissions and increase the satisfaction of residents with public transport. The experience of OECD countries offers lessons for metropolitan governance reforms (OECD, 2015[55]). Recommendations include establishing a regional planning association, adopting digital mobility solutions and integrating housing and transport planning, as argued in the OECD Territorial Review for the Hamburg Metropolitan Region (OECD, 2019[29]). They are complex processes, requiring political support, effective coordination, and reliable funding. The benefits may also take time to materialise. For example, the densification of neighbourhoods well-connected to jobs, services or public transport may take many years and may therefore not be sufficient to reach climate neutrality by 2040.

As an innovative option, on-demand shared mobility services supported by digital technology could also help meet urban mobility needs in a way that eliminates emissions while reducing energy use and harnessing the co-benefits discussed above. Ride-sharing services can be operated on a single integrated platform, where users submit requests and a digital dispatcher matches vehicles to demands on a real-time basis. Considering users' demands and traffic conditions, the system assigns vehicles and generates the optimal routes to destination adhering to pre-set time constraints for all users.

On-demand ride-sharing services available at the doorstep or the next street corner would improve the connectivity of residents with each other and accessibility of jobs or services, especially for low-income households and households in low-density suburban areas, who are often less well connected to public transport. Ideally, all individual car rides in an entire metropolitan area are replaced by shared modes to decrease CO_2 and other negative externalities such as car accidents, air pollution and noise pollution. The advantages of ride-sharing can be maximised when the vehicles are electrified because electric vehicles (EVs) emit less air and noise pollution from engines. Also, the operating cost of electric vehicles is lower than of conventional vehicles, reflecting less repair and higher energy efficiency, a key benefit with intensive use. Users may also benefit more from the development of technology (e.g., battery improvement) as shared vehicles are replaced more often than individual vehicles.

Digital-based ride-sharing can lower CO_2 emissions sharply. In the full replacement scenario where buses and private cars are fully displaced, as modelled for Dublin for example (Box 1.9), vehicle kilometres and CO_2 emissions are substantially reduced, even if internal combustion engines were kept, although the amount of change depends on the size and density of an area, infrastructure and land use (Tennøy and Hagen, 2020[56]). In Ireland's Greater Dublin Area, for example, CO_2 emission can decrease by 42% to baseline if shared modes replace all individual fleets (ITF, 2018[57]). Of course, electrification does away with Scope 1 emissions. However, the estimated emission reductions are still informative on the substantial reduction of energy use. In the case of the Lisbon Metropolitan Area, the estimated impact was more dramatic at 62% (ITF, 2015[58]).

The major co-benefit of mobility-sharing services is decreased road congestion and air pollution. In the Lyon metropolitan area, the lack of public transport and road capacity from the outer part towards the centre to host private car flows are the major sources of traffic jams. Ride-sharing would relieve this issue and increase user satisfaction by decreasing the total number of vehicles on the road. Shared mobility services including car ride-sharing are expected to reduce congestion by 48% according to the Lyon study. In Dublin, the adoption of shared modes in addition to the existing rail and light-rail transit (LRT) can meet local mobility needs with 98% fewer private vehicles in that area. The introduction of ride-sharing services such as Shared Taxis and Taxi-Buses to Dublin is also expected to reduce the total local travel by 38% and traffic jams by 37% (ITF, 2018[57]).

The adoption of ride-sharing services also has the potential to significantly open up public spaces currently occupied by privately owned cars, which tend to remain parked for extended periods. In a simulated scenario for Lisbon, the ITF projected that Shared Taxi and minibus services could lead to a remarkable 95% reduction in the total area allocated for parking (ITF, 2015[58]).

Survey results suggest that 20% of car drivers would be willing to switch to shared rides in Dublin. This share is substantially higher if public awareness is increased with information on how cheaper ride-sharing compares to current private car use. Survey results for Lyon suggest that most citizens are willing to use shared modes.

From the perspective of operators, ride-sharing is also low-cost. For Dublin, the cost of shared minibus services would be less than the price of a public transport ticket, yet would not need to be subsidised. On-demand ride-sharing could be provided at about half the price of public transport offered today. If implemented at a large scale to reduce waiting times, shared mobility would be more desirable for citizens (ITF, 2018[57]). Shared rides could substitute inefficient bus lines, in addition to private car use, and provide feeder service to rail.

Box 1.9. Modelling ride-sharing for Dublin

In the ITF simulation study for Dublin, Shared Taxi and Taxi-Bus replace individual cars and buses while keeping rail and Light-Rail-Transit services. This replacement is modelled on the daily mobility patterns of the Dublin metropolitan area. Shared Taxi is a convenient on-demand service where a maximum of six passengers share a minivan for door-to-door transport. Reservations can be made in real time, and vehicles are assigned to each user adhering to the principle of minimising distance, not only for the requesting user but also for those in the same vehicle. In cases where a Shared Taxi vehicle is unoccupied and not assigned to a new trip, it relocates to the closest depot designated for idle vehicles. Within the model, the fare for a Shared Taxi is estimated to be 75% of the present car cost, ensuring it never falls below EUR 3.

Taxi-Bus service is also on-demand but operates between street corners, using a mini-bus that can accommodate between 8 and 16 passengers. Reservations for Taxi-Bus need to be made 30 minutes in advance. As a shared taxi, Taxi-Bus also follows dynamically optimized routes, ensuring efficient travel between designated stops for all users. The model operates on the assumption that Taxi-Bus fares are set at 50% of the current car cost, with a minimum threshold of EUR 1.5. This calculation is based on a car cost of EUR 10, divided by three trips per day, and added to the fuel cost per kilometre.

Shared Taxi and Taxi-Bus can generate the largest positive impact on reducing congestion and CO_2 emissions when integrated with large-capacity public transport such as light rail services. Both services, operated by professional drivers, offer the option of either a direct, non-stop trip or transport to a rail station if the destination can be reached without transfers (ITF, 2018[57]).

Electric mobility makes road use charging more important

Road use charges will be necessary to replace fossil fuel taxes when fossil-fuel vehicles are phased out, both to replace revenue streams as well as to price negative externalities related to vehicle use such as congestion, accidents, and noise. Road use charges that are time and place-contingent can price externalities more efficiently, especially in urban areas, where external costs are much higher than typical fuel tax rates today (OECD/ITF, 2019[59]).

Since electric vehicles have low operating costs, the diffusion of EVs could intensify car use in cities, aggravating congestion. Automated driving adds to these risks, as it will reduce the opportunity cost of the time spent in the car. Such improvements would lower the cost of private mobility, raising the demand for it, including by encouraging urban sprawl. Large-scale introduction of ride-sharing would also bring efficiency gains and would therefore also need to be accompanied by road charging. Without road use charges, the more efficient mobility services provided by ride-sharing could result in residents increasing mobility demand, for example, by living further away from their workplace, offsetting the effects of ride-sharing on congestion and air pollution.

Road-use charging can include congestion charging (Box 1.10). Policy makers need to decide which roads should be covered. Stockholm has been operating the congestion tax since 2007 to improve transport in central Stockholm. A vehicle is charged every time it passes the charging points, between 6:00 am and 6:29 pm during weekdays. The price depends on the time of day and season, ranging from SEK 11 to SEK 45 each time, with a maximum of SEK 135 per day and vehicle (Transport Styrelsen, 2022[60]). Although public acceptance was less than 40% in 2005, it rose to more than 50% after a trial period in 2006 and to 65% in 2007 after the official adoption. The level of public support further increased to approximately 70% in 2011. Thanks to the charges, traffic fell by 16% in the inner city and by more than 5% outside the cordon, accompanied by a substantial decrease in travel time. In terms of environmental effects, a 10 to 15% reduction in carbon dioxide (CO_2) emissions, a 10 to 14% reduction in air pollutants, and an 8.5% reduction in nitrogen oxides (NO_x) were observed in the inner city. CO_2 emissions in the region of Stockholm decreased by 2 to 3% (Transport Styrelsen, 2022[60]).

Other metropolitan areas of the OECD countries, such as London, Milan and Singapore, have also adopted congestion charges and successfully reduced congestion, travel time, and air pollution (OECD, 2010[61]). In the case of Milan and Singapore, this drop has been linked to vehicle emissions (OECD, 2019[62]). Lessons from the London Congestion Charge show that attitudes change in favour of policies to reduce car demand after their successful introduction as the benefits of less car use materialise (Downing and Ballantyne, 2007[63]).

There are several limitations attached to the development of user charges including the legal ability of subnational governments to create and determine the level of such fees, as well as the capacity and willingness to pay of users (OECD, 2019[62]). Those who can afford all the costs would continue using private vehicles anyway. In other words, such a policy can have negative distributional effects when individuals being taxed do not have alternative means of transport to turn to. Therefore, they need to be accompanied by policies that allow substituting for private car use. public and shared transport systems need to be sufficiently accessible to offset rising inequality as a consequence of a price-based instrument. High-quality ride-sharing could contribute to making more shared transport available.

Box 1.10. Schemes of road-use charging

The most common road-use charging method is the cordon-based system. It charges drivers when they pass charging points, either per day or by the number of crossings. In London, a passenger car that enters the Congestion Charge area at least once a day must pay £15. Auto Pay system bills a driver for the number of charging days the vehicle travels within the Congestion Charge zone, the Low Emission Zone (LEZ) and the Ultra Low Emission Zone (ULEZ) according to vehicle emission standards. There are many other cases of discounts and exemptions. For example, residents of the zone get a 90% discount. Vehicles with nine or more seats that are not registered as buses get a 100% discount. To further reduce cars on the road, electric or hydrogen fuel cell fleets will no longer be eligible for discounts from 2025. In 2023, commemorating the 20th anniversary of this policy, Transport for London reported that the Congestion Charge reduced traffic jams by 30% and promoted walking and cycling by 10%. The office expects that the expansion of the ULEZ to the entire London in August 2023 will bring an additional cutback of 23,000 tCO_2 in outer London, more than the reduction observed in central London. The receipts from the congestion charge are used to finance urban public transport in London (Transport for London, 2023[64]).

An area-based system is another type of road-use charging. It is done by allowing only licence holders to use the roads in charged areas for a certain period covered by the permit. While the cordon-based system can only charge those entering a congestion zone, the area-based system can also cover the trips made within the area. In Singapore, the Area Licensing Scheme (ALS) was introduced in 1975, and drivers had to show a pre-purchased daily pass when entering the downtown priced zone in the morning and evening peak. Cars in shared use and taxis were also subject to the rule, while buses and motorcycles got an exemption. The introduction of ALS reduced 44% of vehicles entering the restricted zone and increased carpooling (8% to 19%) and bus share (33% to 46%). Although both car population and employment in the restricted zone had increased by 1983, the overall auto share for commuters decreased from 56% to 23%. The public transport share in the morning peak time also increased from 33% to 69%. In 1998, the Electronic Road Pricing (ERP) program replaced the ALS. The charges are automatically calculated and collected according to location, time, and vehicle type, using a transponder attached to a vehicle with a pre-paid “smart card” inserted. ERP further reduced 24% of weekday traffic in the restricted zone (U.S. Department of Transportation, 2021[65]).

References

Basner, M. et al. (2014), "Auditory and non-auditory effects of noise on health", *The Lancet*, Vol. 383/9925, pp. 1325-1332, https://doi.org/10.1016/s0140-6736(13)61613-x. [52]

Bishop, K., J. Ketcham and N. Kuminoff (2018), *Hazed and Confused: The Effect of Air Pollution on Dementia*, National Bureau of Economic Research, Cambridge, MA, https://doi.org/10.3386/w24970. [48]

Buck, H. et al. (2023), "Why residual emissions matter right now", *Nature Climate Change*, https://doi.org/10.1038/s41558-022-01592-2. [18]

Chapman, R. (2019), *Managing the transition to a climate-neutral economy in cities and regions*. [8]

Chapman, R. et al. (2018), "A Cost Benefit Analysis of an Active Travel Intervention with Health and Carbon Emission Reduction Benefits", *International Journal of Environmental Research and Public Health*, Vol. 15/5, p. 962, https://doi.org/10.3390/ijerph15050962. [50]

Chen, Z. et al. (2018), "Consumption-based greenhouse gas emissions accounting with capital stock change highlights dynamics of fast-developing countries", *Nature Communications*, Vol. 9/1, https://doi.org/10.1038/s41467-018-05905-y. [16]

City of Rotterdam (2020), *Zero Emission City Logistics Rotterdam*. [37]

City of Seattle (2018), *Seattle Climate Action*. [31]

Climate Science (2022), *Emissions by End-Use: What Activities Create the Most Emissions?*, https://climatescience.org/advanced-emissions-by-product. [68]

Council of Economic Advisers (2014), *The cost of delaying action to stem climate change*. [6]

Dechezleprêtre, A., N. Rivers and B. Stadler (2019), "The economic cost of air pollution: Evidence from Europe", *OECD Economics Department Working Papers*, No. 1584, OECD Publishing, Paris, https://doi.org/10.1787/56119490-en. [42]

Dijkstra, L., H. Poelman and P. Veneri (2019), "The EU-OECD definition of a functional urban area", *OECD Regional Development Working Papers*, No. 2019/11, OECD Publishing, Paris, https://doi.org/10.1787/d58cb34d-en. [35]

Downing, P. and J. Ballantyne (2007), *Tipping Point or Turning Point? Social Marketing and Climate Change*, https://www.ipsos.com/sites/default/files/publication/1970-01/sri_tipping_point_or_turning_point_climate_change.pdf (accessed on 12 October 2023). [63]

EEA (2020), *Environmental noise in Europe 2020*, Publications Office of the European Union, Luxembourg, https://doi.org/10.2800/686249. [51]

Energieswitch (2019), *Rotterdam Climate Agreement*. [32]

Energy Information Administration (2022), *Energy and the environment explained*. [69]

European Commision (2021), *European Emissions Trading System (ETS) – Calculations on the regional employment impact of ETS installations*. [67]

European Commission (2023), *Energy efficiency targets*. [25]

European Commission (2022), *Emissions Database for Global Atmospheric Research (EDGAR)*. [36]

European Commission (2022), *Fit for 55 - Aviation and the EU ETS*, https://climate.ec.europa.eu/eu-action/aviation-and-eu-ets_en. [12]

European Commission (2021), *Aviation and the EU ETS*. [21]

European Commission (2018), *A Clean Planet for all. A European long-term strategic vision for a prosperous, modern, competitive and climate neutral economy*. [3]

Federal Ministry for the Environment, Nature Conservation, Nuclear Safety and Consumer Protection of Germany (2016), *Climate Action Plan 2050*, https://www.bmuv.de/fileadmin/Daten_BMU/Pools/Broschueren/klimaschutzplan_2050_en_bf.pdf. [11]

Frontier Economics (2017), *Affordable Warmth, Clean Growth: Action Plan for a comprehensive Buildings Energy Infrastructure Programme*, https://www.aucklandcouncil.govt.nz/plans-projects-policies-reports-bylaws/our-plans-strategies/unitary-plan/history-unitary-plan/documentssection32reportproposedaup/appendix-3-8-5.pdf. [54]

German Presidency of the Council of the European Union (2020), *EU2020.de*. [20]

Glenk, K. et al. (2021), "The opportunity cost of delaying climate action: Peatland restoration and resilience to climate change", *Global Environmental Change*, Vol. 70, p. 102323, https://doi.org/10.1016/j.gloenvcha.2021.102323. [66]

Grimes, A. et al. (2012), *Cost Benefit Analysis of the Warm Up New Zealand: Heat Smart Programme*, https://www.aucklandcouncil.govt.nz/plans-projects-policies-reports-bylaws/our-plans-strategies/unitary-plan/history-unitary-plan/documentssection32reportproposedaup/appendix-3-8-5.pdf. [53]

Hamburg Chamber of Commerce (2022), . [9]

Hamburg Chamber of Commerce (2010), *Sector Profiles: Commercial Transport in Hamburg*. [30]

Hamburg Ministry for Urban Development and Environment (2019), *First revision of the Hamburg Climate Plan*, https://www.hamburg.de/contentblob/13899086/749a6e50662c96eee81d370f1b0cb631/data/d-first-revision-hamburg-climate-plan.pdf. [72]

Hamburg Statistics Office (2022), *Energiebilanz und CO2-Bilanzen für Hamburg*, https://www.statistik-nord.de/zahlen-fakten/umwelt-energie/energie/dokumentenansicht/product/3381/energie-und-co2-bilanzen-fuer-hamburg-361?cHash=4201529a752424c94a05eb3c4ae751ea. [23]

Hamburger Senat (2022), *Eckpunktepapier für die zweite Fortschreibung des Hamburger Klimaplans*, https://www.hamburg.de/contentblob/16763680/bdac8f8d932cbd784b9256426fc5b11b/data/d-eckpunktepapier2022.pdf. [10]

Heissel, J., C. Persico and D. Simon (2019), *Does Pollution Drive Achievement? The Effect of Traffic Pollution on Academic Performance*, National Bureau of Economic Research, Cambridge, MA, https://doi.org/10.3386/w25489. [49]

Heyes, A., M. Neidell and S. Saberian (2016), *The Effect of Air Pollution on Investor Behavior: Evidence from the S&P 500*, National Bureau of Economic Research, Cambridge, MA, https://doi.org/10.3386/w22753. [43]

H-vision (2017), *Numbers*. [40]

IEA (2022), *CO2 avoidance and use in raw material industries*, https://www.iea.org/policies/13194-co2-avoidance-and-use-in-raw-material-industries. [13]

Intergovernmental Panel on Climate Change (2019), *Climate Change and Land An IPCC Special Report on climate change, desertification, land degradation, sustainable land management, food security, and greenhouse gas fluxes in terrestrial ecosystems.Summary for Policymakers.*, https://www.ipcc.ch/site/assets/uploads/2019/08/Edited-SPM_Approved_Microsite_FINAL.pdf (accessed on 21 October 2019). [27]

Intergovernmental Panel on Climate Change (2018), *Global Warming of 1.5 Degrees Celsius, Summary for policy makers*, Intergovernmental Panel on Climate Change Working Groups I, II, III, Word Metereological Organisation, United Nations Environment Programme, http://report.ipcc.ch/sr15/pdf/sr15_spm_final.pdf. [1]

International Energy Agency (2022), *Energy Efficiency*. [28]

International Energy Agency (2022), *World Energy Outlook 2022*. [24]

International Maritime Organization (2020), *Fourth Greenhouse Gas Study 2020*. [19]

ITF (2018), "Shared Mobility Simulations for Dublin", *International Transport Forum Policy Papers*, No. 58, OECD Publishing, Paris, https://doi.org/10.1787/e7b26d59-en. [57]

ITF (2015), "Urban Mobility System Upgrade: How shared self-driving cars could change city traffic", *International Transport Forum Policy Papers*, No. 6, OECD Publishing, Paris, https://doi.org/10.1787/5jlwvzdk29g5-en. [58]

Karlsson, M., E. Alfredsson and N. Westling (2020), "Climate policy co-benefits: a review", *Climate Policy*, Vol. 20/3, pp. 292-316, https://doi.org/10.1080/14693062.2020.1724070. [44]

Material Economics (2019), *Industrial Transformation 2050 - Pathways to Net-Zero Emissions from EU Heavy Industry*, University of Cambridge Institute for Sustainability Leadership (CISL), https://materialeconomics.com/publications/industrial-transformation-2050. [26]

OECD (2023), *OECD interviews with local stakeholders*. [70]

OECD (2023), *Regional Industrial Transitions to Climate Neutrality*, OECD Regional Development Studies, OECD Publishing, Paris, https://doi.org/10.1787/35247cc7-en. [7]

OECD (2021), *Germany country note to the OECD Regional Outlook 2021 - Progress in the net zero transition*, https://www.oecd.org/regional/RO2021%20Germany.pdf. [71]

OECD (2021), *OECD Regional Outlook 2021: Addressing COVID-19 and Moving to Net Zero Greenhouse Gas Emissions*, OECD Publishing, Paris, https://doi.org/10.1787/17017efe-en. [15]

OECD (2020), *Non-exhaust Particulate Emissions from Road Transport: An Ignored Environmental Policy Challenge*, OECD Publishing, Paris, https://doi.org/10.1787/4a4dc6ca-en. [47]

OECD (2019), "Financing climate objectives in cities and regions to deliver sustainable and inclusive growth", *OECD Environment Policy Papers*, No. 17, OECD Publishing, Paris, https://doi.org/10.1787/ee3ce00b-en. [62]

OECD (2019), *OECD Territorial Reviews: Hamburg Metropolitan Region, Germany*, OECD Territorial Reviews, OECD Publishing, Paris, https://doi.org/10.1787/29afa27f-en. [29]

OECD (2017), *Investing in Climate, Investing in Growth*, OECD Publishing, Paris, https://doi.org/10.1787/9789264273528-en. [4]

OECD (2016), *OECD Territorial Reviews: The Metropolitan Region of Rotterdam-The Hague, Netherlands*, OECD Territorial Reviews, OECD Publishing, Paris, https://doi.org/10.1787/9789264249387-en. [38]

OECD (2015), *Governing the City*, OECD Publishing, Paris, https://doi.org/10.1787/9789264226500-en. [55]

OECD (2010), *Cities and Climate Change*, OECD Publishing, Paris, https://doi.org/10.1787/9789264091375-en. [61]

OECD/ITF (2019), *Tax Revenue Implications of Decarbonising Road Transport: Scenarios for Slovenia*, OECD Publishing, Paris, https://doi.org/10.1787/87b39a2f-en. [59]

Office, H. (1990), *Energiebilanz und CO2-Bilanzen für Hamburg*, https://www.statistik-nord.de/zahlen-fakten/umwelt-energie/energie/dokumentenansicht/product/3381/energie-und-co2-bilanzen-fuer-hamburg-361?cHash=4201529a752424c94a05eb3c4ae751ea. [73]

Port of Rotterdam (2021), *Six infrastructure projects central in first CES Rotterdam-Moerdijk*. [41]

Port of Rotterdam (2018), *The Rotterdam effect: economic significance of the port is twice as high as previously calculated*. [39]

Port of Seattle (2021), *Charting the Course to Zero*. [34]

Sanderson, B. and B. O'Neill (2020), "Assessing the costs of historical inaction on climate change", *Scientific Reports*, Vol. 10/1, https://doi.org/10.1038/s41598-020-66275-4. [5]

Stockholms Stad (2020), *Climate Action Plan 2020-2023*. [33]

Tennøy, A. and O. Hagen (2020), "Reallocation of Road and Street Space in Oslo: Measures for Zero Growth in Urban Traffic", *International Transport Forum Discussion Papers*, No. 2020/14, OECD Publishing, Paris, https://doi.org/10.1787/6d7e9f43-en. [56]

Transport for London (2023), *Congestion Charge marks 20 years of keeping London moving sustainably*, https://tfl.gov.uk/info-for/media/press-releases/2023/february/congestion-charge-marks-20-years-of-keeping-london-moving-sustainably (accessed on 12 October 2023). [64]

Transport Styrelsen (2022), *Congestion tax in Stockholm: Hours and amounts in Stockholm*, https://www.transportstyrelsen.se/en/road/road-tolls/Congestion-taxes-in-Stockholm-and-Goteborg/congestion-tax-in-stockholm/hours-and-amounts-in-stockholm/ (accessed on 11 October 2023). [60]

U.S. Department of Transportation (2021), *Lessons Learned From International Experience in Congestion Pricing*, https://ops.fhwa.dot.gov/publications/fhwahop08047/02summ.htm (accessed on 12 October 2023). [65]

UK Committee on Climate Change (2019), *Net Zero - The UK's contribution to stopping global warming - Committee on Climate Change*, https://www.theccc.org.uk/publication/net-zero-the-uks-contribution-to-stopping-global-warming/ (accessed on 23 October 2019). [2]

United Nations' High-Level Expert Group (2022), *Integrity Matters: Net Zero Commitments by Businesses, Financial Institutions, Cities and Regions*. [22]

van Dender, K. (2019), "Taxing vehicles, fuels, and road use: Opportunities for improving transport tax practice", *OECD Taxation Working Papers*, No. 44, OECD Publishing, Paris, https://doi.org/10.1787/e7f1d771-en. [46]

Wiedmann, T. et al. (2020), "Three-scope carbon emission inventories of global cities", *Journal of Industrial Ecology*, Vol. 25/3, pp. 735-750, https://doi.org/10.1111/jiec.13063. [17]

World Resource Institute, C40 Cities Climate Leadership Group and ICLEI – Local Governments for Sustainability USA (2021), "Global Protocol for Community-Scale Greenhouse Gas Emission Inventories: An Accounting and Reporting Standard for Cities", *World Resources Institute*.. [14]

Xie, Y. et al. (2018), "Co-benefits of climate mitigation on air quality and human health in Asian countries", *Environment International*, Vol. 119, pp. 309-318, https://doi.org/10.1016/j.envint.2018.07.008. [45]

2 Key cross-sector transformations for the transition to climate neutrality for businesses

This chapter analyses key steps businesses can take and the transformations they need to engage to achieve climate neutrality by 2040, across all sectors of the Hamburg economy. It begins by assessing the current progress of Hamburg businesses in implementing net-zero greenhouse gas emission strategies, drawing insights from a survey of selected companies as well as evidence from major companies in Hamburg. This evidence is then compared with international best-practice recommendations. This chapter also assesses specific challenges for SMEs and on the role networks can play to advance SME efforts to decarbonise. Last, this chapter explores how businesses can harness the benefits of low cost renewable energy by adapting to their intermittent nature, as well as urgent action to decarbonise buildings to avoid unnecessary cost.

Businesses, play a critical role to reach net zero emissions by 2050. The transition to climate neutrality brings about material risks for businesses, which translate into financial risks. Climate related risks can be divided into two categories; risks related to the transition to climate neutrality and risks related to the physical impact of climate change (TCFD, 2017[1]). This chapter is only about transition risks. Transition risks include policy actions that constrain emissions, litigation and legal risks, displacement of current technologies requiring fossil fuels, shifts in supply and demand, and reputational risks. In addition to reducing these risks, efforts to mitigate climate change also bring opportunities for businesses. These include resource efficiency and cost savings, development of new products and services, access to new markets, and building resilience along the supply chain.

This chapter assesses key actions businesses need to undertake as well as transformations they need to face to reach climate neutrality by 2040, across all sectors of the economy. It first describes where businesses in Hamburg stand in setting, implementing and disclosing net-zero GHG emissions, drawing on a survey of selected Hamburg businesses as well as on the reporting of Hamburg's biggest companies. It contrasts this evidence with international best-practice recommendations, including from the high-level UN expert group's recommendations for non-state actors, presented in Chapter 1.

The transformations to reach climate neutrality will require integrating new knowledge on business practices that are consistent with climate neutrality, for example on the replacement of fossil fuels or on the integration of business models in climate-neutral value chains. This challenge is particularly big for SMEs who have less resources to invest in such knowledge. It is therefore particularly important for SMEs to work together in networks to benefit from the scale economies of such knowledge. They will also facilitate access to new technologies, infrastructure and needed financing that goes with it. The second section is devoted to this issue.

Reaching climate neutrality requires electrification of most energy use and moving electricity generation to renewables. This is especially true in Germany, which has phased out nuclear energy. Most electricity production from renewables will be based on variable renewable energy (VRE). This includes solar and wind power, sources which vary in the amount of energy provided during the day and across the seasons of the year. This requires business to adopt a new approach to energy use, with opportunities for businesses to take advantage of low-cost VRE and to become more resilient to international energy crises. The third section is devoted to this issue.

Saving energy is another central business challenge to succeed the transformations needed to reach climate neutrality. While this is transversal, covering all activities and asset use, economic impacts will be particularly big in buildings, requiring a major boost to activity notably in employment-intensive construction and installation. Actions to decarbonise buildings and reduce energy use in buildings therefore need to be prepared now and accelerated sharply. The final section is devoted to this issue.

Setting, implementing and disclosing net-zero targets

This section focuses on the climate neutrality or net-zero emission targets of businesses, action plans to reach them and on the assessment of emissions and other metrics that is necessary to measure progress with respect to the targets. The first part of this section sets the scene, reviewing the guidelines and recommendations on setting net-zero targets by businesses. The second part describes where Hamburg businesses stand with respect to targets, emission assessments and action plans, according to available data, including from a survey of Hamburg businesses. The third part discusses regulations that will impact businesses' activities. The last part reviews guidelines for the estimation of Scope 3 emissions, which include emissions in upstream and downstream value chain activities.

To minimize the risks and capture the opportunities in the transitions to climate neutral economies it is essential for businesses to undertake actions to become climate neutral throughout their operations. A survey found that out of 2,000 publicly listed companies, 21% committed to net-zero emissions by 2050 at the latest. The survey also found considerable variation in the quality of business commitments with regards to their scope and implementation plans (OECD, 2021[2]).

There are currently no mandatory requirements for non-state actors to set or report on net-zero GHG emission (or climate neutrality) targets. The UN High-Level Expert Group on Net Zero Emissions Commitments of Non-State Entities offers a roadmap and provides guidance on non-state actors' net-zero targets (Chapter 1). In addition, the European Commission provides guidelines on the disclosure of climate-related risks to inform internal and external stakeholders (Box 2.1), based on work from the Task Force and Climate Related Risks, instituted by the Financial Stability Board in 2017. The OECD Guidelines for Multinational Enterprises on Responsible Business Conduct also emphasize having science-based targets and strategies to achieve net-zero GHG emissions which allow to quantitatively assess progress. This is in line with the UN High-Level Expert Group and the European Commission guidelines.

Box 2.1. European Commission guidelines on disclosure

1. **Business Models**

It is important for internal and external stakeholders, notably the board, investors, and customers, to understand the company's view of how climate change impacts its business model and strategy and how its activities affect the climate, over the short, medium, and long run. For this, companies need to take a longer-term perspective than they normally do for financial reporting. The companies need to describe climate-related risks and opportunities for the company's business model, strategy and financial planning. The companies should describe how the company's business model can impact the climate, both positively and negatively. They could also describe the resilience of the company's business model and strategy, taking into consideration different climate related scenarios over different time horizons.

2. **Policies and due diligence processes**

Companies need to describe their policies related to climate, including mitigation and adaptation policies. They should describe climate-related targets, notably GHG emission targets, and how their targets relate to national and international targets. They should describe how the company is engaging with its upstream and downstream partners. They need to clarify the board's oversight of climate related risks and opportunities and describe the management's role in assessing and managing climate related risks and opportunities.

3. **Outcomes**

Companies should describe the outcomes of the company's policy on climate change, including the performance of the company against targets set to manage climate-related risks and opportunities, in particular the development of GHG emissions.

4. **Principal risk and management**

Companies should describe how they identify, assess, and manage climate related risks over the short, medium, and long term. They should describe the processes for prioritising climate related risks, including any thresholds applied and indicate which risks across the value chains are considered most significant. They need to disclose how scenarios or carbon pricing are used for risk management actions.

5. **Key performance indicators**
   - **a.** Companies need to disclose key performance indicators relevant to their business. Companies need to provide a description of and any changes in the methodologies used to calculate or estimate the indicators. They can consider disclosing the following indicators:
     - **i.** *GHG emissions* → Scope 1, 2 and 3 emissions and GHG absolute emission targets
     - **ii.** *Energy* → total energy consumption and production from renewable and non-renewable sources, renewable energy consumption and/or production target
     - **iii.** *Physical risks* → assets in regions likely to become more exposed to physical climate risks
     - **iv.** *Products and services* → % of turnover (in the reporting year) from products or services associated with activities that meet criteria for contributing to mitigation or adaptation to climate change.
     - **v.** *Green finance* → Climate-related green bond ratios (green bonds outstanding divided by total bonds outstanding) as well as any green debt ratios.
   - **b.** In addition to the above-mentioned indicators, companies can also consider the following:
     - **i.** Sector-specific indicators relevant to the industry. Use the TCFD's guidance.
       - **a)** TCFD guidelines provide sector-specific guidance for energy, transportation, materials and buildings, and agriculture, food, and forest products
     - **ii.** Indicators related to natural capital
     - **iii.** Indicators related to human capital and social issues
     - **iv.** Indicators related to opportunities (revenues from low-carbon products, revenues from product or services applying to the circular economy model, and R&D expenditures in circular economy production).

Source: European Commission, Guidelines on reporting climate-related information.

The majority of Hamburg businesses still need to set a net-zero target

This subsection draws on a survey of businesses in the Hamburg Chamber of Commerce (HCC), conducted by the HCC, to provide insights on their net-zero GHG emission commitments, their perceptions of related challenges and opportunities, as well as progress in addressing them. It also draws on an evaluation ten biggest companies' reporting.

The survey was sent to a sample that is broadly representative of the population of Hamburg businesses in the sectors covered by the HCC, although businesses with fewer than 10 employees are underrepresented. The questionnaire was sent to a random sample of 2300 businesses with more than 10 employees and to a random sample of 500 businesses of all size classes. 128 businesses responded to the questionnaire (the number of respondents varies slightly across questions), a response rate of about 4%. This response rate is similar to other business surveys conducted by the HCC. Businesses that are relatively strongly engaged with the climate policy agenda are more likely to respond. This needs to be taken into account when interpreting the survey results reported below.

The sectoral composition of responding businesses corresponds, in broad orders of magnitude, to the sectoral composition of the members in the HCC (Annex Table 2.A.1). Service businesses are somewhat underrepresented, except in logistics, transport as well as in media and ICT. Industrial and construction businesses are slightly overrepresented. However, the size distribution of responding businesses differs strongly from the population of businesses. Businesses with less than 3 employees are heavily underrepresented. All other firm sizes are overrepresented, especially the biggest ones.

As the results below show, unsurprisingly, small firms are much less likely to adopt climate neutrality objectives. Service companies, which are also somewhat underrepresented, also appear to have a smaller propensity to adopt emissions objectives. This also suggests that the share of business with climate targets is overall overestimated. However, larger firms, which are more likely to have climate neutrality targets, also have a bigger weight in Hamburg's economy and employment.

Low response rates among small businesses suggest they do not have sufficient capacity to invest in needed knowledge. This point is taken up below in the section on business networks for SMEs to build climate-neutral business models.

The questions posed in the survey can be found in the Annex.

Results from the questionnaire

66% of responding businesses report that climate protection is important or very important for them. They see substantial opportunities in reaching climate neutrality. In particular it boosts the employer's branding attractiveness (Figure 2.1). Indeed, most of the responding businesses see a climate neutrality target as an opportunity to improve their attractiveness as an employer and in product marketing.

Figure 2.1. The biggest perceived opportunity from climate neutrality is employer attractiveness

Answers to the question: What business opportunities do you see for your company in the field of climate neutrality?

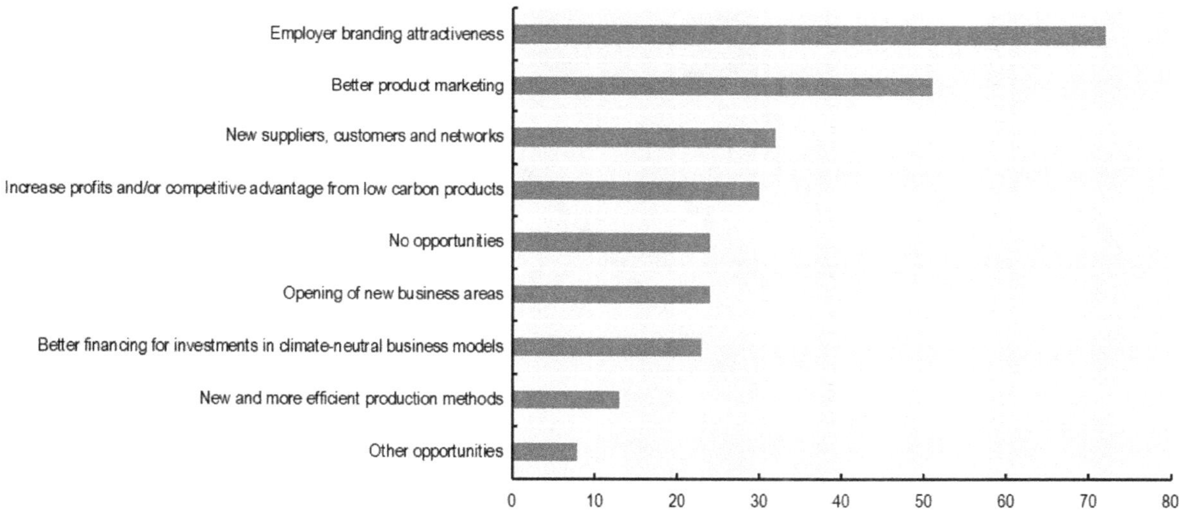

Note: Total of responding businesses is 128. Multiple answers were possible.
Source: Survey carried out by HCC (2023).

StatLink https://stat.link/1ox63s

58 responding businesses (48%) have set a net-zero emissions target (Annex Figure 2.A.1). Among these 76% have set targets for 2040 or earlier, in line with the HCC's climate neutrality target and around 70% have set intermediate targets. Among the responding businesses, lacking resources to do so was the most frequent reason not to have a climate neutrality target (Annex Figure 2.A.2). Most of the businesses with net-zero targets include their Scope 1 and Scope 2 emissions in these targets, but only 45% include Scope 3 emissions. 25 (41%) businesses have reported the net zero target publicly, 19 (31%) have not but are planning to, and 14 have not. Most of the companies have defined responsibilities in management and reduced their energy consumption. However, other measures such as investment in climate neutral technologies, circular economy or climate neutral products in sales or purchasing are still in

implementation, or in planning (Figure 2.2). For the phase-out of fossil fuels and the scale-up of renewables a majority of all responding businesses state they have set a target for at least either of the two, although few respondents (around 10%) have a target to exit fossil fuels.

Figure 2.2. Defining management-level responsibilities and reducing energy consumption have been implemented by a majority of respondents

Answers to the question: What measures does your company plan to undertake to achieve climate neutrality?

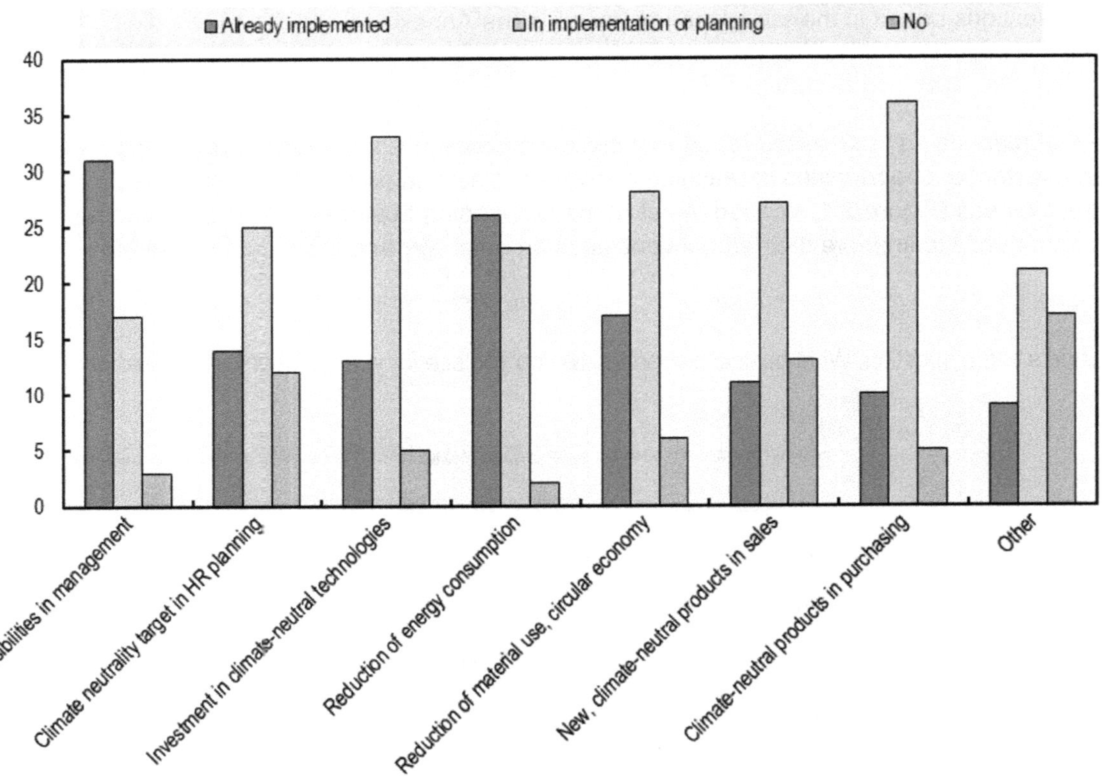

Note: Total number of responding businesses are those with a climate neutrality target (58). Multiple answers were possible.
Source: Survey carried out by HCC (2023)

StatLink https://stat.link/exz42m

Businesses in information and communication technology as well as in the finance sector appear to be among the most likely to be ambitious, setting net-zero targets of 2040 or earlier, judging by survey responses (Figure 2.3). However, the number of firms responding in each sector is small. The most ambitious companies tend to be medium-sized. The majority of small companies have yet to set a net-zero target.

Figure 2.3. Medium-size, ICT and finance businesses set the most ambitious net-zero targets

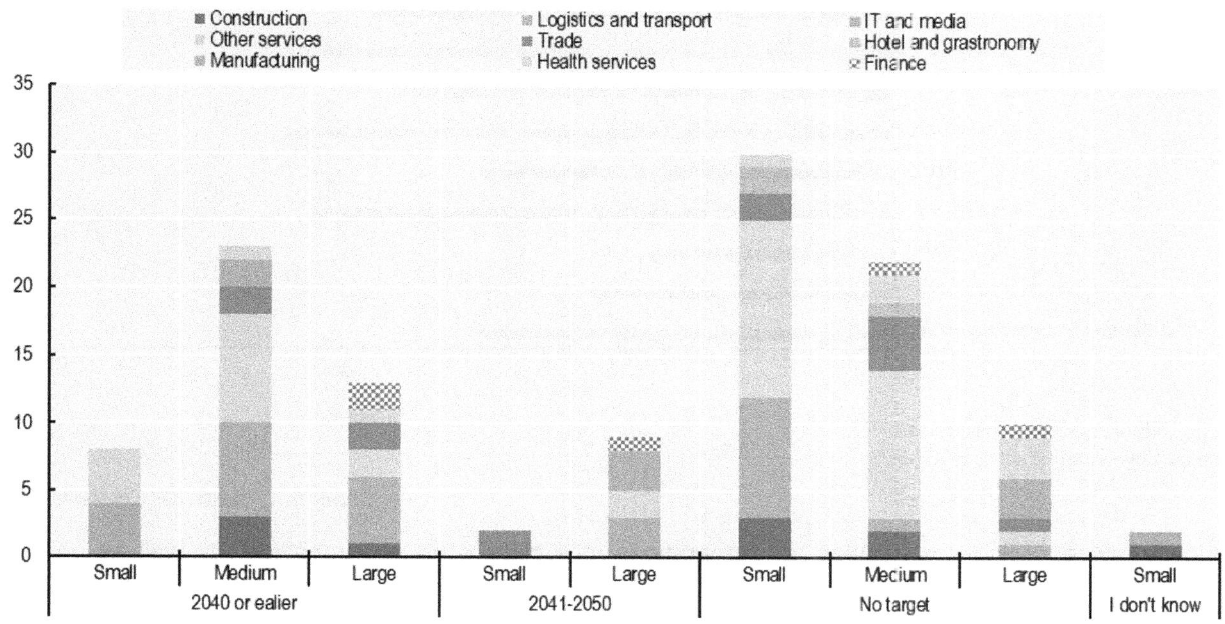

Note: Total of responding businesses is 128 and number of responding businesses with a climate neutrality target is 58. The classification of business size is the following: Small for companies with up to 49 employees, medium for companies with 50-249 employees and large for companies with 250 employees or more.
Source: Survey carried out by HCC (2023).

StatLink https://stat.link/r50m1l

Only 38% of businesses reported that they measure and disclose any relevant indicators. Scope 1 and 2 emissions are the indicators the most monitored and disclosed. Only 30 businesses (23%) monitor Scope 3 emissions, of which one half also disclose them (Figure 2.4). Moreover, most companies don't monitor energy efficiency, ecosystems impact of supply chains and revenues from low carbon products when these are monitored.

Figure 2.4. Scope 1 and Scope 2 emissions are the most monitored and disclosed key indicators for the transformations to reach climate neutrality

Answers to the question: Which climate indicators does your company monitor?

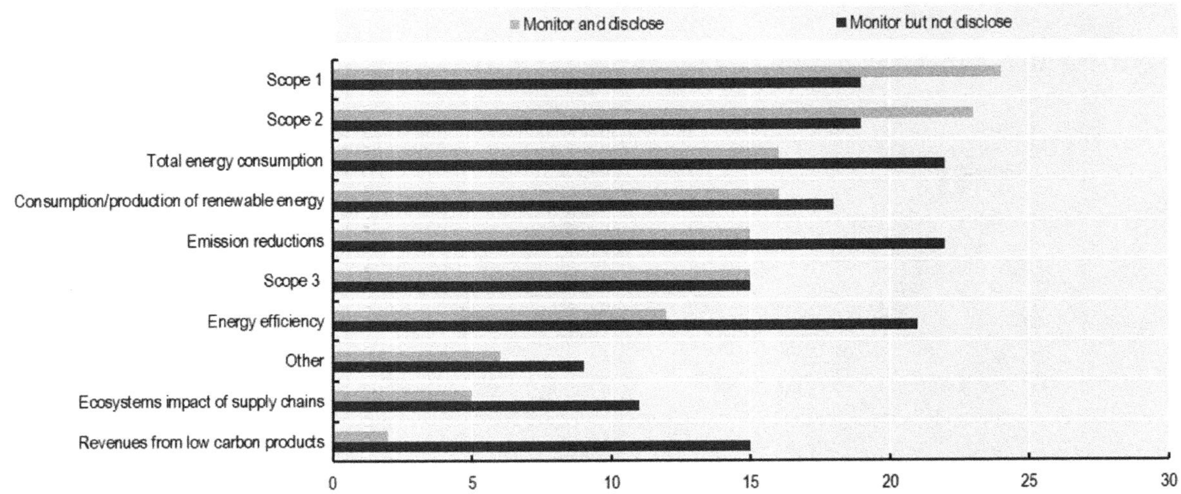

Note: Total of responding businesses is 128.
Source: Survey carried out by HCC (2023).

StatLink https://stat.link/saih8x

Uncertainty regarding costs, sales, and operations in a climate neutral transition, as well as needed investments in climate-neutral production and their financing are identified the most often as challenges in the transition to climate neutrality. Higher production costs and a lack of qualified workers are also of concern (Figure 2.5).

Figure 2.5. Most Hamburg business perceive uncertain prospects and investment funding as a challenge of the climate neutrality transition

Answers to the question: What business challenges do you see for your company in the field of climate neutrality?

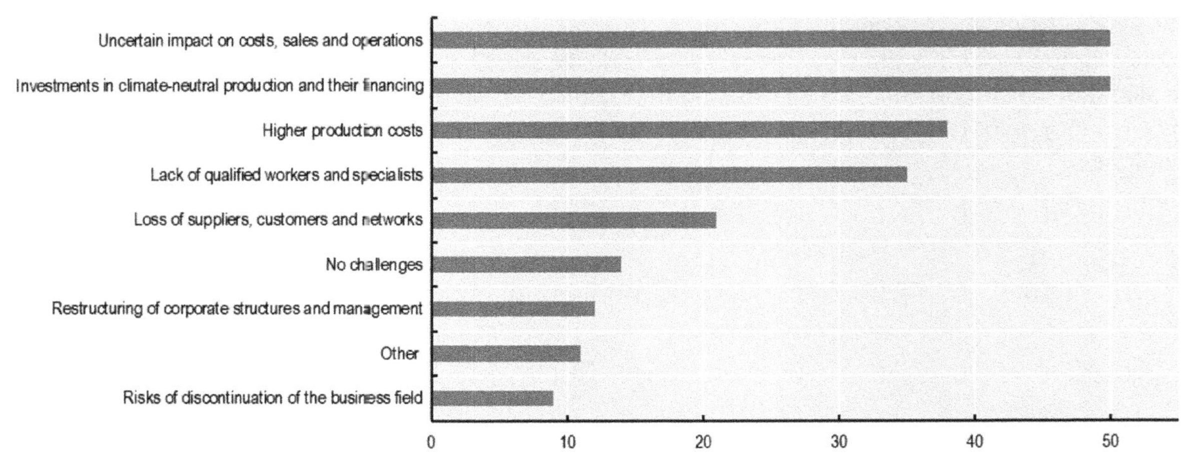

Note: Total of responding businesses is 128.
Source: Survey carried out by HCC (2023).

StatLink https://stat.link/tjw7a0

Most businesses lack financial support. By contrast, advisory bodies are available sufficiently. However, all type of support needs to be stepped up to meet the responding businesses' expectations (Figure 2.6).

Figure 2.6. Most businesses feel that financial support is available but insufficient

Answers to the question: How would you evaluate the availability of the following support services on the way to climate neutrality?

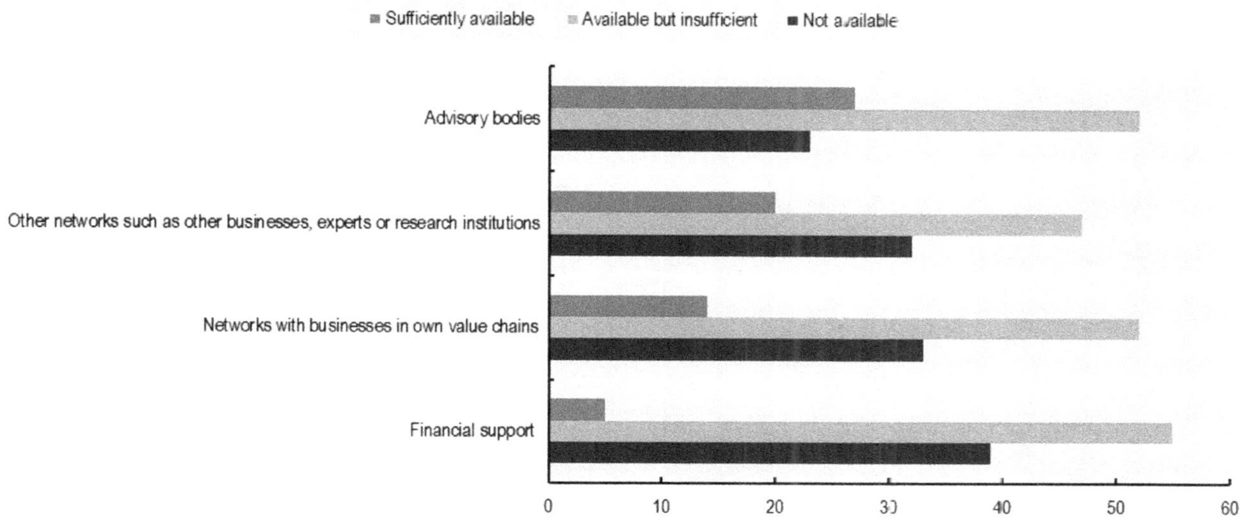

Note: Total of responding businesses is 128.
Source: Survey carried out by HCC (2023).

StatLink https://stat.link/49i3jl

Where do the biggest companies in Hamburg stand?

The biggest 8 companies in Hamburg, according to turnover, have set and disclosed net-zero and intermediate targets (Table 2.1). However, only some include Scope 3 emissions in their net zero and intermediate targets. As argued in chapter 1, Scope 3 emissions are important to assess climate-related risks and opportunities. The companies with net-zero targets for 2050 need to update their ambition to be in line with Germany's climate neutrality target for 2045 and the 2040 HCC target. The company with the earliest climate neutrality target is Vattenfall GmbH, with a 2040 net-zero target covering Scope 1 and Scope 2 emission. There is a wide range of intermediate targets in terms of what emissions are included, and how emissions are measured. Furthermore, among the biggest 8 companies, climate targets of Aurubis AG, Vattenfall GmbH, Airbus Operations GmbH, and Beiersdorf AG obtained certification for being aligned with the 1.5°C scenario from the Science Based Target initiative (SBTi). The SBTi is an independent organisation that provides climate target advice. The Otto Group and Nordex have adopted a Science Based Target (SBT), which is currently being validated by the Science Based Target initiative (SBTi). The Otto Group intends to follow this target from the 2024 financial year onwards. Five companies are implementing a decarbonisation action plan.

Table 2.1. Net-zero and intermediate targets from the top 8 companies in Hamburg by turnover

Company	Turnover (million €)	Emissions (tonnes of CO_2 eq)	Net-Zero Target	Intermediate Targets	International Offsets	Action Plans
Hapag-Lloyd	22274	**Scope 1** - 13,390,000 **Scope 2** – 9,000 **Scope 3** – 2,302,000	**2045** - Net zero GHG emissions for the fleet	**2030** - reduction of 30% emission intensity compared to 2019		Available
Aurubis AG	17064	**Scope 1 + 2** - 1,605,000 **Scope 3** - 6,181,000	**2050** - Carbon neutrality (latest)	**2030** - reduction of 50% of Scope 1 + Scope 2 emissions from 2018 **2030** - 24% reduction of Scope 3 emissions per ton of copper cathodes from 2018		In development
Otto GmbH & Co KG	16200		**2045** – Net zero	**2025** - reduction in 40% GHG emissions compared to 2018 2030 – carbon neutral in all locations, transport and employee mobility	Offsetting mainly used for the carbon neutral shipment of parcels	Not available
Vattenfall GmbH	13970	**Scope 1** – 9,500,000 **Scope 2** – 200,000 **Scope 3** – 16,100,000	**2040** – Net zero (Scope 1 + Scope 2)	**2030** – reduction of 77% of Scope 1 + Scope 2 emission intensity compared to 2017 **2030** – Reduction of 33% of absolute Scope 3 of sold products compared to 2017	Offset of all the business travel emissions through CO2 certificates in the UN's Clean Development Mechanism system	Available
Airbus Operations GmbH	8396	**Scope 1 + 2** – 857,000 **Scope 3** - 505,643,000	**2050** – Net zero (Scope 1 + Scope 2) and support decarbonization of the industry (Scope 3)	**2030** – reduction in 63% Scope 1 + Scope 2 GHG emissions compared 2015 **2030** – reduction in 46% Scope 3 GHG emission intensity generated by commercial aircraft	Remove 100% of residual emissions by 2030 through offsets	Available
Beiersdorf AG	7600	**Scope 1** – 90,349 **Scope 2** – 1,320 **Scope 3** - 1,034,279	**2050** – Net zero (at the latest)	**2025** – 30% absolute reduction in Scope 1,2,3 GHG emissions compared to 2018 **2030** – 100% climate neutral production	Offset flight emissions. Aim to offset remaining GHG emissions.	In development
ArcelorMittal GmbH	7400	**Scope 1 + 2** – 145,800,000	**2050** – Net zero	**2030** – 35% reduction in Scope 1 + Scope 2 emission intensity in Europe compared to 2018	5% of total emissions will be offset by buying high-quality offsets	Available
Nordex SE	5694	**Scope 1** – 27,164 **Scope 2** – 501 In production sites as well as main offices **Scope 3** - 3,483,000	Committed to set net-zero targets in line with the SBTi	No intermediate targets announced	Working to identify suitable carbon offset projects to compensate for remaining Scope 1 and Scope 2 emissions by the end of 2023.	Available

Note: Scope 3 emissions only relates to Consumer Business.
Source: (Hapag-Lloyd, 2022[3]) (Aurubis, 2023[4]) (Otto Group, 2022[5]) (Vattenfall, 2022[6]) (Maxingvest AG, 2022[7]) (Marquad & Bahls, 2020[8]) (Beiersdorf, 2022[9]) (Airbus, 2022[10]) (ArcelorMittal, 2021[11]) (Nordex, 2022[12])

Companies should prioritise measures to reduce emissions over the purchase of carbon credit to compensate for emissions. Only residual emissions which are unavoidable owing to technological or financial limitations may be compensated. Most of the 8 biggest enterprises in Hamburg are relying on international offsets, although to varying extent. Some companies such as Airbus Operations GmbH and ArcelorMittal GmbH are planning to adopt Carbon Capture Storage (CCS) technology.

Do Hamburg businesses follow the recommendations and guidelines?

Businesses need to scale up their ambition to be aligned with the recommendations from the UN High-Level expert group and the European Commission. Most businesses surveyed have yet to set net-zero targets or publicly announced them. Out of the businesses who have set net-zero targets only a minority include Scope 3 emissions in those targets. In some of Hamburg's 8 biggest companies' targets it is not clear which of the Scope 1 and Scope 2 emissions are included. Some of the top ten biggest companies include carbon offsets as means to mitigate residual emissions. Even so, the carbon offsets all seem to be of high quality and in line with the conditions of additionality and permanence sets by the High-level expert group (Chapter 1).

In terms of disclosure, a small number of the surveyed businesses monitors and discloses relevant climate indicators, and even out of the 8 biggest companies, not all disclose emission numbers for all 3 Scopes. Monitoring progress, using these and other metrics, is essential for the businesses to make informed decisions on their path to climate neutrality.

Regulations businesses could expect

Businesses in Hamburg should expect policies at the EU and the German Federal level which will require businesses to strengthen the integration of environmental commitments in the operations and supply chains (OECD, 2021[13]). A few examples of recent policies are the following:

- **EU Mandatory Due Diligence Rules:** Companies will be required to identify, prevent, end or mitigate adverse impacts on human rights and the environment. Once the proposal is adopted by the European Parliament and the Council, Members States will have two years to transpose the requirements into national law. Companies will be required to assess the impact of their business activities on climate change, as well as the impact of their full supply chains on the environment. This includes identifying greenhouse gas emissions, assessing the risks of climate change to their operations, and taking action to mitigate those risks (European Commission, 2022[14]).
    - **German Due Diligence Act:** From 2023, companies based in Germany with more than 3,000 employees must ensure that social and environmental standards are observed in their supply chain. Companies must set up processes to identify, assess, prevent, and remedy human rights and environmental risks and impacts in their supply chains and in their own operations. These processes must be published in their annual reports. From 2024, this will apply to companies with more than 1,000 employees (Sedex, 2023[15]).
- **EU Regulation on Deforestation-free Supply Chains:** Once adopted, goods in the EU market must no longer contribute to deforestation and forest degradation globally. All relevant companies will have to conduct strict due diligence of their supply chains that use any products of palm oil, cattle, soy, coffee, cocoa, timber, and rubber, as well as any derived products. Companies will be required to collect precise geographical information on where their commodities are grown and sources. Micro and small enterprises will have a longer adaptation period, but this will apply to all companies (European Commission, 2022[16]).
    - **German Raw Materials Strategy:** The strategy aims to ensure long-term security of supply of raw materials needed in industrial production in Germany and ensure socially and environmentally fair supply chains. The strategy focuses on raw materials sourcing, material and resource efficiency, sustainability and transparency and international cooperation. The

strategy is in line with European policies and initiatives on raw materials and circular economy (International Energy Agency, 2022[17]).

- **EU Carbon Border Adjustment Mechanism (CBAM)**: In October 2023, the CBAM will enter into its transitional phase, initially focusing on the most carbon-intensive's imported goods, in particular cement, iron, steel, aluminium, fertilisers, as well as electricity and hydrogen. Businesses with these imported goods in their supply chains will have to report embedded GHG emissions (both direct and indirect emissions in the supply chain). From January 2026, importers will need to surrender the corresponding tradable CBAM certificates, for which the price will be calculated depending on the weekly EU ETS price (European Commission, 2022[18]).
- **The Renewable Energy Directive II** defines criteria for sustainable biofuels production. From 2023, this regulation is mandatory for a defined list of stakeholders. If the criteria are not respected, the company couldn't claim to contribute to reach a specific target, couldn't receive public subsidies or can have their subsidies frozen.

Assessing scope 3 emissions

To be aware of business model opportunities and risks businesses need to broadly measure their Scope 3 emissions (Box 2.2). Commonly used reporting standards provide some guidance, but challenges remain as reporting approaches are numerous and differ from each other. Businesses need to strike a balance: the accuracy of assessment and reporting on each emission source should depend on the importance for the business model, taking into account feasibility (World Economic Forum, OECD, Business at OECD, 2023[19]). Interoperability of reporting across firms can help improve clarity and reduce cost. The Greenhouse Gas Protocol, a collaboration between the Word Resources Institute and the Carbon Trust, offers technical guidance for calculating Scope 3 emissions in corporate value chains (Greenhouse Gas Protocol, 2013[20]).

Box 2.2. Scope 3 emissions and their categories

Scope 3 emissions can be divided into upstream and downstream emissions, including 15 different categories (Figure 2.7).

- **Upstream emissions** include all the indirect emissions during the production of a good or service. These include the extraction, production and transport of goods and services purchased by the company, any extraction, production and transport of fuel and energy purchased by the company, any transport and distribution of intermediary products, any waste generated during the operations, emissions from business travel and employee commuting, and any emissions from upstream leased assets.
- **Downstream emissions** occur during the use or disposal of a company's product, the consumption and end of life period. These include emissions from any transport and distribution of products sold, processing of sold products, use of sold products, end-of-life treatment of sold products, downstream leased assets, franchises, and investments.

Figure 2.7. Scope 3 emissions in corporate supply chains and their 15 categories

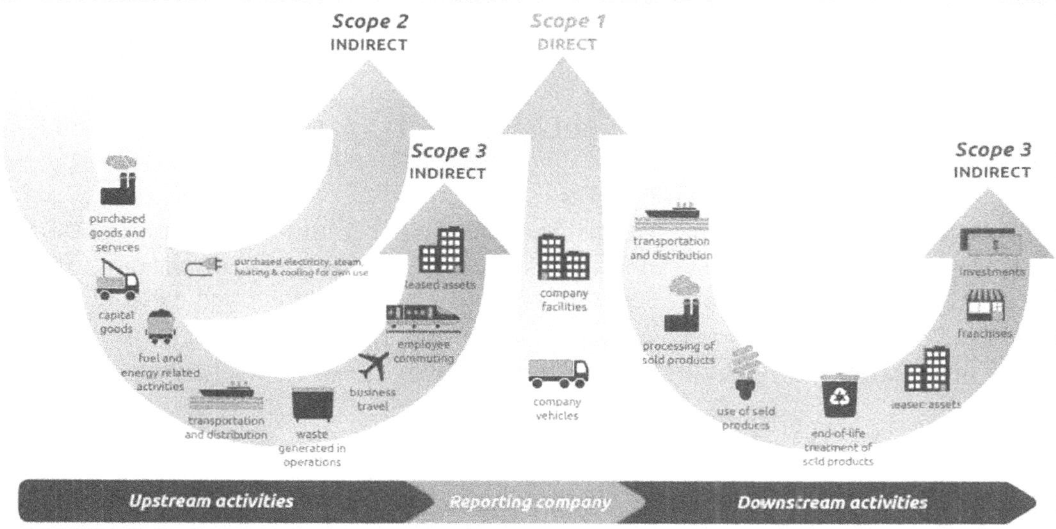

Source: (Greenhouse Gas Protocol, 2013[20])

The protocol recommends that businesses identify which Scope 3 emissions are a priority to calculate or estimate. Businesses need to start by understanding which emissions are the most relevant according to some suggested criteria:

- Emissions with the **highest reduction potential**
- Emissions that pose **great risk to the company**
- Emissions of great **importance to key stakeholders**
- Emissions that are **highly relevant in the sector**
- Emissions that require **high level of spending to abate**
- Emissions of which abatement **generate high level of revenue**

Once the most relevant Scope 3 emissions are identified, the protocol recommends how to measure or estimate those emissions. The protocol provides calculation and estimation guidance for each of the categories (Figure 2.7). It provides businesses with guidance on which calculation method is best suitable given available company resources. The guide outlines the data needed, the emission factors needed, as well as data collection guidance

The data businesses can use to calculate Scope 3 emissions are either primary or secondary. **Primary data** refers to emissions from specific activities within a business' value chain, including provider-specific energy use, or emissions data provided by suppliers. **Secondary data** are industry-averages from published databases, government statistics, industry associations, financial and proxy data. For secondary data, companies can use internationally recognized or peer-reviewed databases. The protocol provides secondary data sources that businesses can consider when calculating Scope 3 emissions (Box 2.3).

Box 2.3. Overview of secondary data sources for estimation of Scope 3 emissions

The data sources suggested by the Greenhouse Gas Protocol meet two criteria; they offer direct access to users without need to purchase specific software tools and provide online resources for users to review documentation and additional information. Some of the data sources listed as examples are the following:

- **The European Aluminium Association (EAA)** provides up to date life cycle inventory (LCI) for aluminium production and transformation processes in Europe.
- **The European Copper Institute (ECI)** provides life cycle assessments (LCA) of three types of copper products: tubes, sheets and wires.
- **The Global LCA Data Access (GLAD)** a coalition of datasets providers, provides a global network of LCA databases that connects multiple data sources to support LCA in a way that facilitates sustainability-related decisions.
- **Greenhouse gases, Regulated Emissions, and Energy use in Transportation (GREET)** provides energy and emission impacts of advanced vehicle technologies and new transportation fuels, the fuel cycle from wells to wheels and the vehicle cycle through material recovery and vehicle disposal.
- **The International Iron and Steel Institute (IISI)** supplies LCIs of fourteen steel industry products.
- **The IPCC Emissions Factor Database** is a library of emission factors and other parameters with background documentation and technical references.
- **LCI calculation tools for Regionalised Waste Treatment** allow users to generate LCI datasets for treatment of solid waste.
- **Scope 3 Evaluator,** a global standard in value chain GHG accounting, allows users to estimate emissions for all 15 Scope 3 categories. It helps companies take the first steps in measuring and reporting Scope 3 emissions. Companies can use this information to start identifying areas in which to pursue more accurate inventory and focus their reduction efforts.

Source: (Greenhouse Gas Protocol, 2023[21])

The protocol then provides decision trees for each Scope 3 emission category to help identify the calculation method. Figure 2.8 illustrates the decision tree for emissions arising from purchases of goods and services.

Figure 2.8. Decision tree for selecting a calculation method for emissions from purchased goods and services

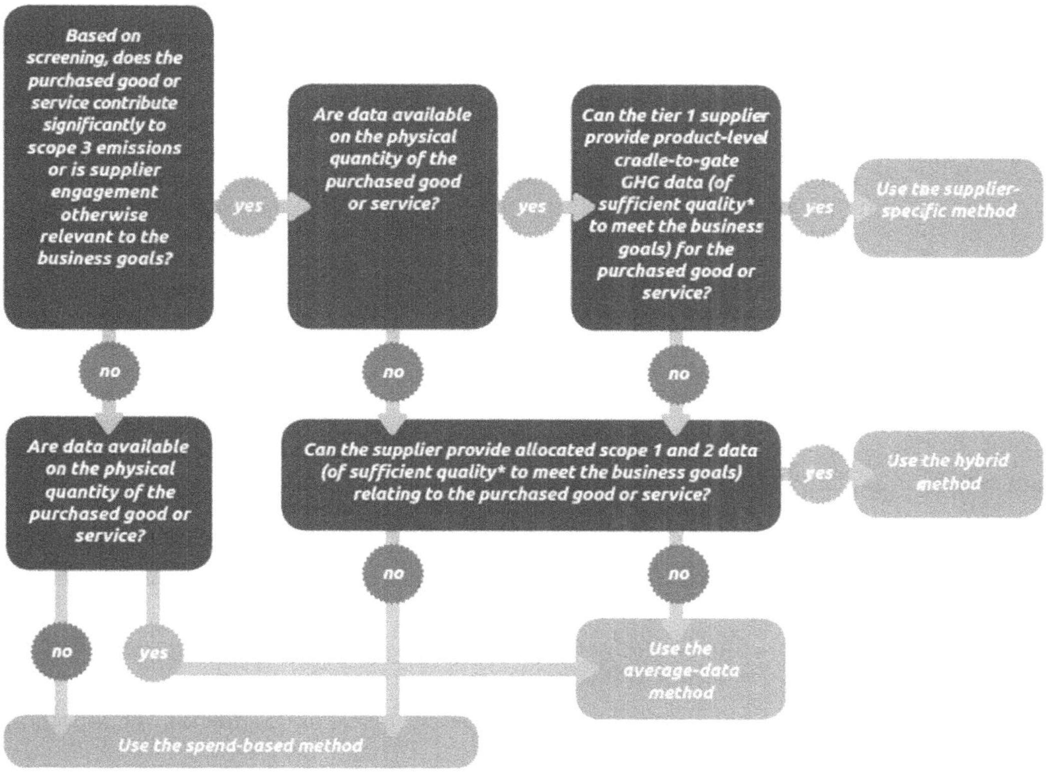

Source: (Greenhouse Gas Protocol, 2013[20])

The methods for the calculation and estimation of Scope 3 emissions include (Table 2.2):

- **The supplier-specific method** collects product-level cradle-to-gate GHG inventory data from goods and services suppliers.
- **The hybrid method** uses a combination of the supplier-specific method and secondary data to fill the gaps. This includes calculating upstream emissions of goods and services from suppliers' activity data for the materials, fuel, electricity used, from distance of transportation or waste generation from the production of goods and services.
- **The average-data method** estimates emissions for goods and services by collecting data on mass or other relevant units of goods and services purchased and multiplying by the relevant secondary emission factors.
- **The spend-based method** estimates emissions for goods and services by collecting data on the currency value of goods and services purchased and multiplying it by relevant secondary emissions factors.

Table 2.2. Calculation methods and data requirements for Scope 3 emissions of purchased goods and services

Calculation method	Activity data needed	Emission factor needed
Supplier-specific method	• Quantities or units of goods or services • purchased	• Supplier specific emission factors for the purchased goods or services from the suppliers
Hybrid method	• Allocated scope 1 and 2 data by supplier relating to the good or service purchased by the reporting company. • Mass or quantity of material inputs used by supplier to produce purchased goods. • Mass or quantity of fuel inputs used by supplier to produce purchased goods. Distance from the origin of the raw material inputs to the supplier. • Quantities of waste output by supplier to produce purchased goods. • Other emissions emitted in provision of the purchased goods as applicable	Depending what activity data has been collected from the supplier, companies may need to collect: • Cradle-to-gate emission factors for materials used by tier 1 supplier to produce purchased goods. • Life cycle emission factors for fuel used by incoming transport of input materials to tier 1 supplier. • Emission factors for waste outputs by tier 1 supplier to produce purchased goods. The secondary emission factors required will also depend on what data is available: • Cradle-to-gate emission factors of the purchased goods or services per unit of mass or unit • Cradle-to-gate emission factors of the purchased goods or services per unit of economic value
Average data method	• Mass or number of units purchased of goods or services for a given year	• Cradle-to-gate emission factors of the purchased goods or services per unit of mass or unit of product
Spend-based method	• Amount spent on purchased goods or services, by product type, using market values	• Cradle-to-gate emission factors of the purchased goods or services per unit of economic value

Source: (Greenhouse Gas Protocol, 2013[20])

Standards in reporting of Scope 3 emissions

Once Scope 3 emissions have been estimated, the protocol provides guidelines for reporting them. A business is advised to:

- Report Scope 3 emissions separately from Scope 1 and Scope 2 emissions
- Report Scope 3 emissions separately for each category
- Report in CO_2 equivalents
- Report the types and sources of data used, and a description of data quality
- Report the method used to calculate Scope 3 emissions

Among the 8 biggest companies in Hamburg not all companies that report Scope 3 emissions are reporting the types of data, data quality and the methodology. For example, Hapag-Lloyd was preparing Scope 3 accounts to be available in the second half of 2023 in its sustainability report.

Table 2.3. Top 8 biggest companies and Scope 3 emissions

Company	Report Scope 3 separately from Scope 1 and Scope 2	Report categories of Scope 3 emissions separately	Report in CO_2 equivalents	Report the types, sources of data, and data quality	Report methodology
Hapag-Lloyd	YES	YES	YES	-	-
Aurubis AG	YES	YES	YES	YES	-
Otto GmbH & Co	-	-	-	-	-
Vattenfall GmbH	YES	YES	YES	YES	YES
Airbus Operations GmbH	YES	YES	YES	YES	YES
Beiersdorf AG	YES	YES	YES	YES	YES
ArcelorMittal GmbH	-	-	-	-	-
Nordex SE	YES	YES	YES	YES	YES

Source: (Hapag-Lloyd, 2022[3]) (Aurubis, 2023[4]) (Otto Group, 2022[5]) (Vattenfall, 2022[6]) (Maxingvest AG, 2022[7]) (Marquad & Bahls, 2020[8]) (Beiersdorf, 2022[9]) (Airbus, 2022[10]) (ArcelorMittal, 2021[11]) (Nordex, 2022[12])

Key actions

Immediate action:

- Businesses need to set a net-zero emissions target, with clear intermediary targets in intervals of, for example 5 years, for at least Scope 1, Scope 2 emissions.
- Businesses need to assess Scope 1 and 2 emissions. They should begin assessing Scope 3 emissions as well as related climate risks and opportunities.
- Businesses need to prepare an action plan to reach the net zero target by 2040.
- The HCC can provide guidance on these steps as well as working groups to facilitate knowledge-sharing and coordination.
- Businesses should invest in just transition efforts, for example all businesses with operations in developing countries should demonstrate how their net zero transition plans contribute to the economic development of regions they are operating in.

By 2030

- On the basis of Scope 3 emission estimates, business need to work towards reaching climate neutrality in their supply chains following science-based emission reduction scenarios consistent with the Paris Agreement.
- Businesses with reporting requirements need to publicly report their GHG emissions and other relevant metrics, such as energy use, and the methods employed. They should disclose net zero targets, intermediate targets and the action plan to meet their net zero target, including actions for reducing Scope 3 emissions in line with the Paris agreement. They need to report on their progress against intermediate targets. Disclosures need to be accurate and reliable. Businesses can also seek independent evaluation of their annual progress reporting and disclosures.
- The HCC can create a public platform to monitor progress on these actions, provide guidance, facilitate knowledge-sharing and coordination.
- The HCC can provide guidance on all these steps as well as working groups to facilitate knowledge-sharing and coordination.

By 2040

- Businesses need to reduce their own Scope 1 and 2 emissions to zero leaving residual positive emissions until 2050 only in a few activities hard to decarbonise. These can be offset with the purchase of carbon credits.

- Businesses and the HCC need to find high-quality certified offsets for residual emissions.
- The HCC can provide guidance on these steps as well as working groups to facilitate knowledge-sharing and coordination.

Networks for SMEs to build climate-neutral business models

Small and medium sized enterprises (SMEs)[1] are key actors to achieve net zero emissions. In OECD countries, SMEs represent at least 50% of greenhouse gas (GHG) emissions and 30-60% of energy use in business sectors (OECD, 2022[22]). Besides, the flexible, risk-taking profile of SMEs makes them essential players in generating and adopting green innovations. Developing clean innovations involves a high level of risk that large companies are often unwilling to take. To illustrate, in the UK, SMEs account for around 90% of clean technology companies (OECD, 2022[22]).

Hamburg's economic landscape includes a large number of SMEs, especially SMEs with fewer than 50 employees. They are particularly numerous in the wholesale and retail trade sector (Figure 2.9). SMEs in this sector are likely to share challenges to make business models consistent with climate neutrality, for example, with respect to assessing and reducing Scope 3 emissions. Some of them may have strong links to the port. SMEs are also heavily involved in transport, real estate activities and construction. Transport includes some of the activities that are most difficult to decarbonise, notably in freight (Chapter 3), while construction firms will make an important contribution to the decarbonisation of buildings. SMEs involved in some professional, scientific, and technical activities can participate in the creation and dissemination of green technologies. This section argues that SMEs may have much to gain from cooperating in networks to address these challenges collectively.

Figure 2.9. SMEs are primarily found within the sectors of trade and business services

Sectoral contributions to the number of businesses in each business size class, in per cent

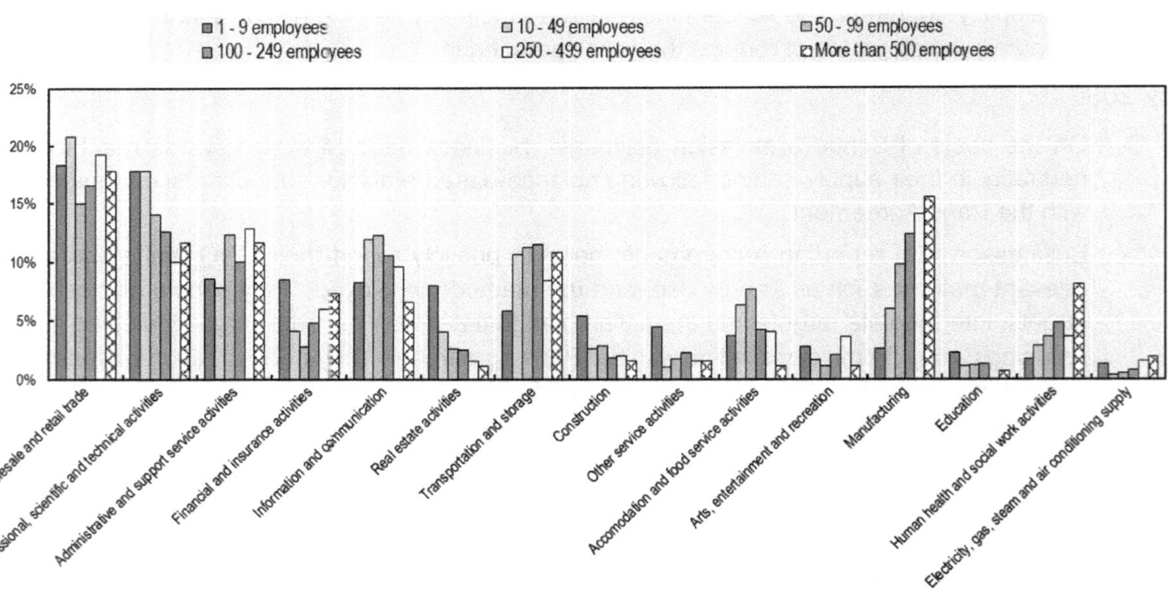

Note: Sectors representing less than 1% of companies in any category are not taken into account.
Source: Hamburg Statistics Office (2022)

SMEs often lack the resources needed for the transformations to reach climate neutrality. They have more limited access to new technologies, information, and finance, which prevents them from investing in the technologies needed for decarbonisation. In 2021, the UK Chamber of Commerce conducted a survey showing that only 10% of SMEs measure their greenhouse gas emissions, while a Korean survey shows that 31% believe that the main obstacle is a lack of information on the methods to use (OECD, 2022[22]). Given their limited resources, market uncertainty and risk may affect SMEs more, making them more vulnerable. In Korea, 60% of SMEs believe that the net-zero target will reduce their competitiveness. In the Hamburg metropolitan region, SMEs experience more difficulties in innovating and adopting digital technologies than in other regions (OECD, 2019[23]).

Box 2.4. Cluster networks in Hamburg

The approach adopted by the City of Hamburg centers on industrial clusters. As part of Hamburg's "*InnovationsAllianz*," the city collaborates with stakeholders from diverse sectors, including science, business, politics, institutions, and associations, to identify key activities for cluster development. Each cluster is coordinated through a central management system overseen by local public authorities and is funded through a combination of membership fees and public support. They regularly provide policy recommendations.

Hamburg's strategic clusters, in line with the Smart Specialisation Strategy, cover a range of sectors such as "Life Sciences," "Logistics," "Aviation," "Media & IT," "Renewable Energies," "Creative Industries," "Finance," and "Maritime Industries." Furthermore, a new food cluster is set to be established by the end of 2023, adding another significant sector to Hamburg's economic landscape.

The cluster agencies do not have specific climate neutrality objectives themselves but contribute to the implementation of general or sectoral requirements through their projects. Noteworthy contributors to these efforts include, but are not limited to, agencies such as Renewable Energy Hamburg (EEHH), Hamburg Aviation (HAv), Logistics Initiative Hamburg (LIHH), and Maritime Cluster Northern Germany (MCN).

Participating in a network can ease SMEs' transition towards climate neutrality, pooling knowledge and resources. These networks typically fall into four categories: production/supply chain networks, knowledge/innovation networks, and strategic partnerships or clusters (Figure 2.10; (OECD, 2023[24])). Networks offer an efficient avenue for accessing knowledge, allowing to harness economies of scale in the use of such knowledge. Furthermore, network participation can help mitigate uncertainties. Networks can also facilitate access to financial resources. Additionally, SMEs aiming to establish themselves as environmentally responsible enterprises can seek membership in networks that uphold high sustainability standards, to market business models providing climate-neutral solutions. Figure 2.6 suggests that many Hamburg companies consider such networks are not sufficiently available. The HCC should encourage their development. Some cluster initiatives have been developed by the regional government (Box 2.4).

Figure 2.10. Typology of SME networks

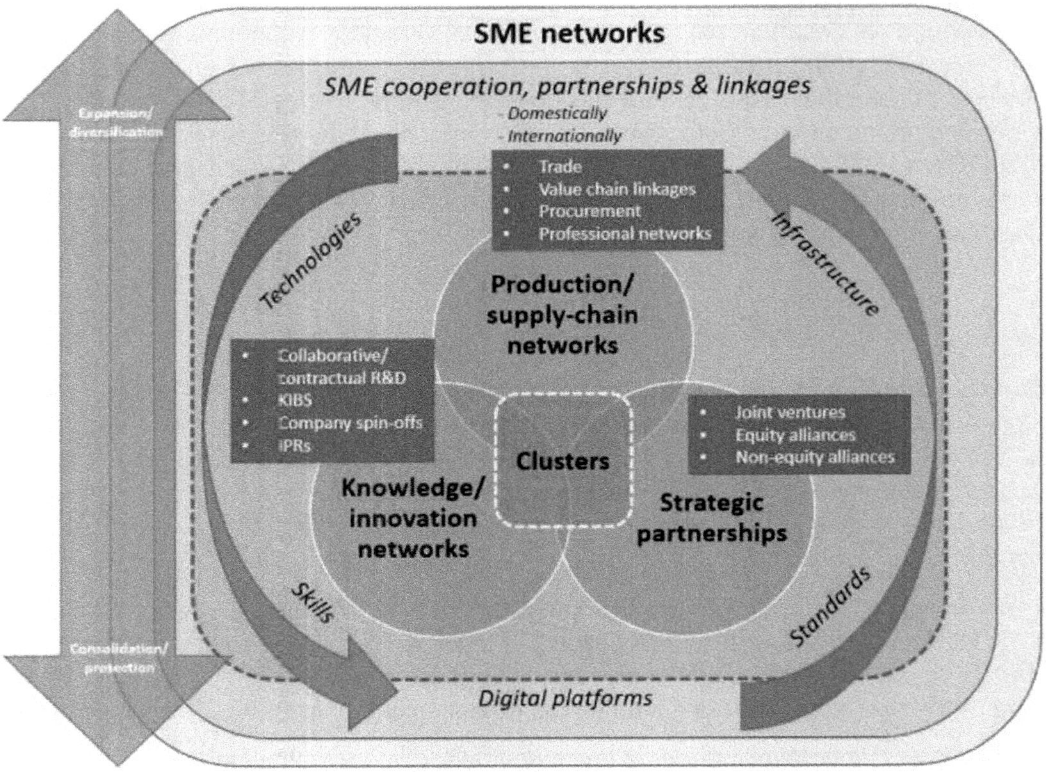

Source: (OECD, 2023[24])

Box 2.5. Business networks can meet a range of SME resource needs for the transition to climate neutrality

A network connects businesses with each other, enabling transactions that cannot be intermediated through markets, including knowledge. In the case of SMEs networks, the entities are companies, and the transferable flows can be technologies, data, behaviours, and tacit knowledge (know-how). There are four channels through which flows are transferred:

- **Supply-chain networks:** connect firms' activities involved in the production of a good or service. Actors (e.g., suppliers, competitors, or multinationals) are connected through trade, investment, professional networks, digital platforms, or networking facilities. They exchange products, services, financial flows, innovations spill over and intangible assets. As part of the net-zero carbon transition, a large company could provide operational and financial support to its SME suppliers to enable them to use greener technologies. For instance, the Apple's Supplier Clean Energy Program is a program designed to help Apple suppliers decarbonize their production. Through this program, Apple works closely with its suppliers to identify and implement solutions to reduce their energy consumption and carbon footprint.
- **Knowledge networks**: collaborative framework to develop and share knowledge and innovation. The entities involved are innovative SMEs, High Education Institutions (HEI), Public Research Institutions (PRI), government and intermediaries. Those actors share codified and tacit knowledge, R&D, data, skills, technology, financing, and intangibles through contractual or collaborative R&D, consultancy, Knowledge-intensive business services (KIBS), training, labour mobility, patenting and licensing, spin-off, digital platforms, and networking facilities. An example of a knowledge network to accelerate the transition to a carbon-free economy is the EIT Climate-KIC, which is a European Knowledge and Innovation Community (KIC). They bring together several green innovation players, including companies, universities, and public institutions. In addition, the EIT Climate-KIC raises and invests funds, develops educational programs to enhance environmental skills, and has an incubation program.
- **Strategic partnerships:** formal collaboration between two or more entities (e.g., start-ups, multinationals, universities, PRIs) to develop motivation for new products and/or commercialisation. They can also exchange infrastructures through R&D joint ventures, research consortium, join R&D agreements and minority holdings, licensing, or franchising. Yulex Corporation is an example of an SME that has created a strategic partnership with a larger company, Patagonia, to participate in the decarbonisation of the economy. They collaborated in the development and use of plant-based materials for wetsuits instead of synthetic neoprene, which is highly polluting.
- **Clusters:** interconnected entities such as companies, HEIs, research institutes and technology providers, bridging institutions or customers. Cluster can be organized around a geographical area or sectoral proximity. To illustrate, the Sustainable Packaging Cluster in Catalonia, Spain is a cluster program that facilitates connections between SMEs and other players to develop environmentally friendly packaging and support the diffusion of green packaging technologies.

Networks are a key factor in the decarbonization of SMEs

- Knowledge and innovation networks are essential for the creation and diffusion of the innovation SMEs need for climate neutral business models. Small companies that cooperate are more innovative than large companies that do not (Figure 2.11). Knowledge and innovation networks can connect public institutions, companies, High Education Institutions (HEI) or Public Research Institutions (PRI) to exchange knowledge, expertise, and new technologies.

Figure 2.11. Innovative firms cooperate more than non-innovative ones

Share of companies that cooperate with other companies or organisations as part of their commercial activities by size class, among non-innovative and innovative companies (2020).

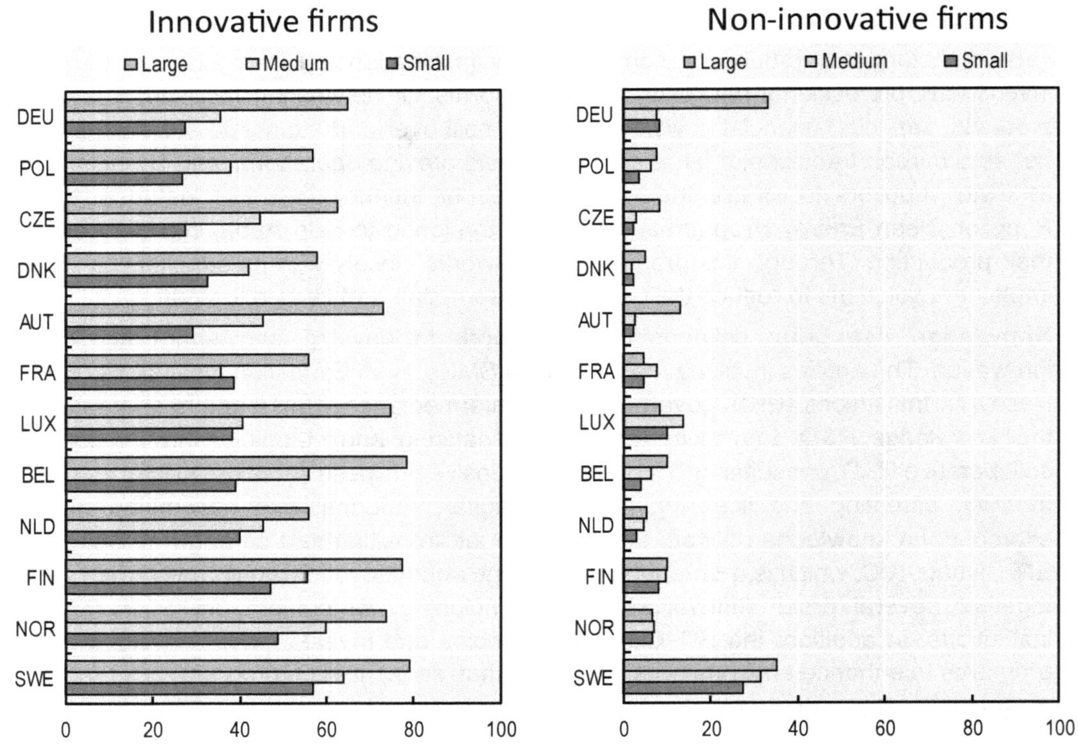

Source: (OECD, 2023[24])

Knowledge networks help SMEs adopt climate-neutral business practices at lower cost. The knowledge required for developing innovation is increasingly specialized and vast, increasing the cost of acquiring it. Joining a network to gain access to this knowledge is a solution for SMEs wishing to innovate while sharing the cost. Knowledge networks enable SMEs to pool data or physical infrastructure, enabling significant economies of scale. More generally, belonging to a network comprising KIBS players gives SMEs access not only to scientific and technological services, but also to legal, accounting and management services, engineering, R&D and IT systems, as well as design and advertising services, among others. These types of services are not at the heart of SME activities, but they are essential to their survival and to the dynamism of the economic fabric.

Networks can facilitate the access to finance in several ways. Many knowledge networks help SMEs to identify and raise private and public funding, some even having their own funds. For instance, EIT Climate KIC raises and invests funds to finance green projects developed by SMEs (Box 2.5). Venture capitalists and business angel investors can be key partners for start-ups thanks to their management and strategic resources, but they are sometimes hard to reach. Participation in network programs offers the opportunity to meet investors. Finally, through strategic partnerships, SMEs can be financed by larger companies.

Access to data at business level are particularly key. Networks can help by providing access to the technology and sources. SMEs need to identify areas for improvement. Measuring electricity consumption, for example, helps to understand electricity requirements.

This knowledge can also be transferred via the supply-chain network, through buyer-supplier or parent-subsidiary relationships or strategic partnerships. Larger firms with more resources could help SMEs to gain information about existing green technologies and best practices, and to implement them, securing net-zero value chains. For instance, in the Aligned Business Framework Program Ford encourages and helps its suppliers to implement energy-efficient technologies.

Through supply chain or strategic partnership networks, SMEs can gain insights into their supply chain operations and how to decarbonize them. Large companies must comply with increasingly stringent environmental regulations. They are also increasingly scrutinized and considered responsible for the entire production chain. As a result, multinationals are paying increasing attention to the environmental performance of the SMEs they deal with. For example, following the adoption of the Corporate Sustainability Reporting Directive (CSRD), several large European companies cancelled contracts with SMEs that had failed to submit sustainability performance reports.

Networks can also serve to set environmental standards and identify members as fulfilling them for marketing purposes. Participating in such networks enables SMEs to gain more information on the environmental standards to be achieved and to identify themselves as complying with those standards, including on emissions reduction. Identification as a "green" business is important for reputation and helps gain visibility with clients, investors, workers and consumers who need access to environmentally high-performing SMEs.

Key actions

Immediate action:

- The HCC could identify the extent to which Hamburg SMEs are integrated in networks to help undertake transformations needed to reach climate neutrality.
- The HCC could support the creation of networks to help undertake transformations needed to reach climate neutrality, including academics, research institutions, SMEs and large companies to exchange information on technologies and best practices. Networks could provide an online platform and regular meetings/events.
- Businesses, with the support of HCC, should initiate the creation of networks on strategic issues identified in this report such as coordinated decarbonization across value chains, cost savings from better renewable energy use (this chapter); infrastructure needs from the electrification of road freight (Chapter 3) or the circular economy (Chapter 4).

Making better use of low-cost renewable energy

This section shows how Hamburg companies can benefit from low-cost electricity produced with variable renewable energies (VRE) by adapting to their intermittent nature. Germany has set an ambitious target of 80% renewable energy sources by 2030 (OECD, 2023[25]), and the IEA forecasts a 50% increase in electricity demand in a net-zero emissions scenario by 2050 (IEA, 2022[26]). Taking advantage of low-cost electricity from renewables when it is abundant will become increasingly important, as renewable electricity use needs to expand rapidly to substitute for fossil fuels.

There is a business case for variable renewable energies (VRE)

VRE have the lowest electricity production costs, even when incorporating battery storage costs (Box 2.6). VRE are expected to play a significant role in Germany, particularly in Northern Germany, which possesses potential to emerge as a leader in renewables with high onshore wind energy production in the Schleswig Holstein region and offshore wind in the North Sea. Collaborative initiatives, such as the Norddeutsches Reallabor (NRL) initiative, have been initiated. The North German Energy Transition initiative, launched between 2016 and 2020, brings together a coalition of 60 public, private, and research stakeholders spanning the two federal states of Hamburg and Schleswig-Holstein. The aim of the initiative is to meet energy demand by providing entirely renewable energy in both states by 2035 (OECD, 2019[23]). However, the existing high-tension power line network is presently insufficient to transport a substantial amount of renewable electricity to Hamburg. This limitation could result in increased electricity prices for the city, even when wind power is abundant. Local photovoltaic (PV) systems, especially solar rooftop panels, emerge as a promising and cost-effective solution for Hamburg, to avoid such bottlenecks. In 2022, Hamburg's total annual electricity demand stood at 10.4 terawatt-hour (TWh), whereas the estimated potential for PV generation reached 6.37 TWh (Hamburg Klimabeirat, 2023[27]). Currently, solar rooftop adoption in Hamburg is low. Installed capacity of organic photovoltaics in Hamburg stands at 77 megawatt peak (MWp), in contrast to the 190 MWp in Berlin.

Table 2.4. Solar rooftop adoption in Hamburg is lower than in other Länder

Installed capacity of organic photovoltaics (OPV) in selected Länder

Länder	Installed capacity (MWp)	Installed capacity by inhabitant (kWp/Inhab)
Berlin	190	61
Bremen	66	110
Hamburg	77	44

Notes: MWp stands for megawatt peak and kWp for kilowatt peak. This is a unit of measurement of the potential power of sources such as solar or wind energy, which takes account of variations linked to factors such as the intensity of sunshine or wind speed.
Source: (Hamburg Klimabeirat, 2023[27])

Box 2.6. Electricity from VREs is cheaper than fossil fuel fired electricity

The levelized cost of electricity generation is the lowest for VRE and should continue to fall, widening the gap with fossil fuel-fired electricity (NEA/IEA, 2020[28]). Among VRE, utility-scale solar panels produce the cheapest electricity, although onshore wind turbines remain very competitive. In Germany, photovoltaic (PV) technologies and offshore wind power plants are expected to see the biggest drop in cost (Kost et al., 2021[29]).

Figure 2.12. The cost of electricity generation is the lowest for solar and wind

The levelized cost of electricity of renewable energy technology and conventional power plants in Germany in 2021

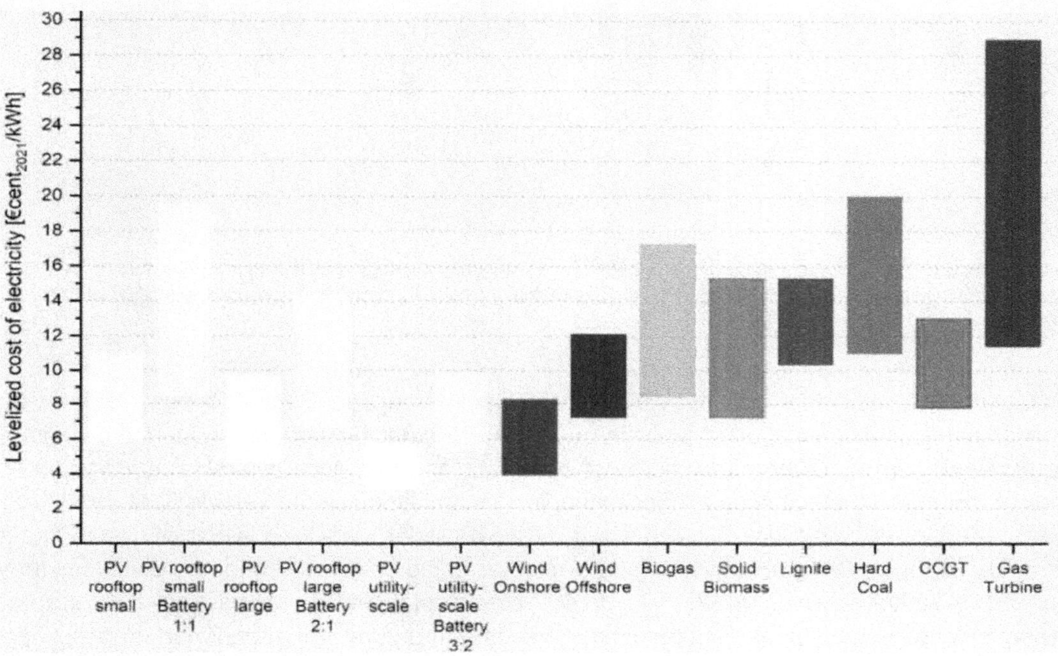

Notes: CCGT= combined cycle gas turbines
Source: (Kost et al., 2021[29])

Renewable energy production may not align with current electricity demand patterns. For instance, solar energy generation is highest during the daytime and in the summer, whereas electricity demand typically peaks in the evenings and during the winter months. The International Energy Agency (IEA) anticipates a significant rise in the demand for electricity system flexibility within the European Union by 2030, driven by an increased share of renewables (Figure 2.12).

Figure 2.13. Hour-to-hour flexibility needs rise significantly by 2030 in major markets

Hour-to-hour flexibility needs in the United States, European Union, China and India in 2021 and assuming that all targets announced by government are met on time and in full for 2030

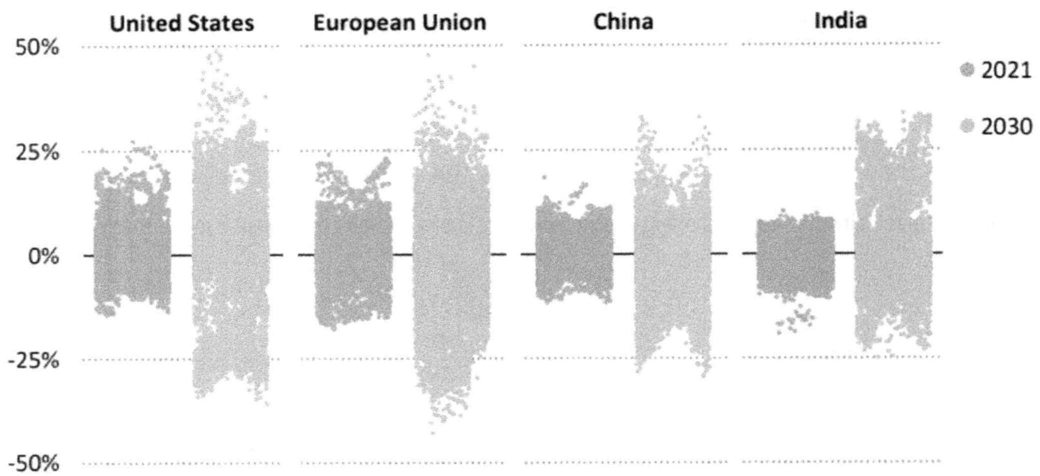

Note: This corresponds to the Annouced Pledges Scenario (APS). Flexibility needs are the needed hour-to-hour rate of change in electricity demand (Change in Demand / Time Period) needed to match hourly wind and solar PV production to hourly electricity demand.
Source: (IEA, 2022[26])

Business can make fuller use of low-cost VRE by using sector coupling. Sector coupling refers to the closer integration of energy-using economic activities with VRE electricity production, to make optimal use of available renewable energy when it is the most abundant and therefore available at lowest cost, taking advantage of the needed electrification. For example, electrification of the vehicle fleet can be combined with smart charging infrastructure. Vehicles with large batteries can feed electricity back into the grid (Hossain et al., 2016[30]). Companies with electric vehicles could be rewarded for the electricity they supply in peak periods (Vanholme et al., 2022[31]).To do so, firms must invest in vehicles that are capable to store and release electricity and in bidirectional charging infrastructure. In a transport hub like Hamburg, efficiently managing an electric vehicle fleet can be a particularly valuable tool.

For activities difficult to electrify such as heavy industry or shipping, sector coupling can also involve converting renewable electricity in other storable energy carriers, a concept known as power-to-X technology (Erbach, 2019[32]). For instance, power-to-gas refers to the conversion of the surplus electricity to hydrogen or synthetic gaseous fuels (Gea-Bermúdez et al., 2021[33]), which could be used by Hamburg's industry, notably steel and chemicals production. Power-to-heat technology refers to the conversion of electricity to heat with, for instance heat-pump.

Sector coupling generates collective benefits. Indeed, it increases capacity utilization in renewable electricity production and improves grid efficiency, thus reducing decarbonization costs from infrastructure development, while speeding up the transition to climate neutrality. For Germany, a sector coupling scenario compared to a fully electric scenario would generate an annual cost saving of €12 bn by 2050 (Van Nuffel et al., 2018[34]), around 15% of annual spending on electricity in recent years. VRE coupled with sector coupling reduces exposure to fuel price volatility and political risk from fossil fuels (Krane and Idel, 2021[35]). Hamburg's role as a transportation hub, coupled with the expected high demand for hydrogen and the as yet undeveloped potential of rooftop solar panels, suggests that sector coupling could result in substantial cost savings for the city.

Individual companies using sector coupling also draw individual benefits (Van Nuffel et al., 2018[34]). First, sector coupling improves energy efficiency. According, to the IEA, charging vehicles during the day when electricity demand is low could decrease charging cost from €0,39 to €0,28 per kWh (IEA, 2023[36]), once prices respond to fluctuations in supply and demand. Additionally, investing in sector coupling creates synergies, as equipment for energy end use can also be used to store and release electricity to save costs. For example, electric vehicles can be used for transport or to store electricity (Box 2.8). With Hamburg hosting numerous firms in the road freight sector, where electrification with battery electric vehicles should be expected to accelerate, sector coupling presents a compelling opportunity. Sector coupling can also raise the benefits to businesses from installing rooftop solar panels, boosting their use of self-generated electricity.

Encouraging sector-coupling through enhanced flexibility in the electricity market

Policy makers should enhance electricity market flexibility to encourage firms to develop sector coupling. Higher granularity of the electricity market helps better reflect the variable supply of VRE over time and across locations and encourages investment in flexibility, including through sector coupling (Kraftnät, 2017[37]). Currently, the wholesale pricing timing (imbalance settlement period) in Germany stands at 15 minutes, but there is potential to further reduce it to align with VRE patterns. Transitioning to a pricing structure in smaller, subnational zones or nodes can enhance the accuracy of price signals, strengthening the efficiency of energy markets, with favourable impacts on competitiveness. Northern Germany, and Hamburg in particular, would be able to put to good value industry and port related activities (Chapter 3) and derive advantages from implementing geographically tailored electricity pricing, which could result in lower prices (IEA, 2022[38]).

Demand response programs can encourage firms to adapt electricity consumption according to wholesale price signals. Policy makers and energy utilities have a role in making these programs available and clear. Incentive based programs reward market participants for adapting to VRE supply patterns through direct payments. Electricity demand can also be managed by utilities directly (Box 2.7). In price-based programs, such as Time of Use programs, prices are higher at peak times (Albadi and El-Saadany, 2008[39]).

Hamburg offers demand response programs in line with national initiatives. These programs include interruptible load programs, auctions, and opportunities for renewable energy producers to engage in direct marketing. These tend to benefit large customers. Demand response aggregators could also assist small and medium-sized consumers in participating collectively. Several initiatives provide real-time data and automated control for electricity consumption. To encourage more demand response, it is crucial to have clear regulations, and information for all participants. One of the main obstacles to developing demand response is the absence of standardized processes and contracts governing settlements between aggregators (consumers) intermediaries and suppliers (ENEFIRST, 2020[40]).

To be able to make their electricity demand more flexible, companies need to invest in digital tools such as demand monitoring and demand response technologies. Digital technologies such as smart meters, energy management systems and automated control systems can be used to closely monitor, control, and forecast energy consumption (IEA, 2022[38]).

Box 2.7. Demand response program types

Incentive Based Programs

IBP programs reward market participants for their adaptation to VRE supply patterns through direct payments.

1. **Direct load Control:** Utilities could remotely manage and control participant equipment (e.g., air conditioners and water heaters) in response to variations in wholesale electricity prices, reflecting energy supply conditions. In exchange, customers receive a financial incentive.
2. **Interruptible/Curtail Programs:** Customers receive upfront incentive payments or rate discounts if they agree to reduce their demand to predefined values in periods of high grid stress.
3. **Demand Bidding**: Customers bid on specific load reductions and if the bid is accepted, the customer must reduce their load by the amount specified in the offer.
4. **Emergency DR Programs:** Customers are paid incentives for measured demand reductions during emergency conditions.
5. **Capacity Market:** Customers commit to providing pre-specified demand reductions in case of peak demand in the long term. They are penalized if they fail to respond to calls.
6. **Ancillary services market:** Customers bid for discounts/payments for offering load reduction on the spot market as an operating reserve. When bids are accepted, participants are paid the spot market price for committing to standby and are paid the spot market energy price if their load curtailments are required.

Price Based Programs

In price-based programs, price signal or tariff are used to change energy demand.

1. **Time of Use:** The price of electricity per unit of consumption differs according to time of day. Prices are higher at peak times and lower at off-peak times.
2. **Critical Peak Pricing / Extreme Day Pricing:** In case of contingencies, a higher electricity price is paid in exchange for a lower price on normal days.
3. **Real Time Pricing:** Customers pay the real time wholesale electricity price.

Source: (Albadi and El-Saadany, 2008[39])

Solar PV rooftop systems can lower energy costs

Solar PV rooftop systems offer a particularly appealing prospect for businesses in Hamburg to generate and consume their own electricity to benefit from low-cost electricity, ultimate contributing to developing a decentralized system based on distributed energy resources (Hargroves et al., 2023[41]). In the absence of granular electricity pricing over time, sector-coupling can offer opportunities for businesses to benefit more from self-production/consumption of VRE, as self-produced electricity tends to be cheaper than electricity purchased from utilities. The production time profile of solar panels, primarily during daylight hours, corresponds well with the electricity consumption patterns of businesses. When the production does not match the demand of electricity, it is possible to couple self-production/consumption with battery or thermal storage (Figure 2.14; (Renewable Energy Agency, 2021[42])). Even in the case of low electricity needs, self-production/consumption with sector coupling is attractive. In 2023 the German government has increased feed-in tariffs for distributed solar PV (IEA, 2022[43]). Finally, it is important to note that Hamburg's photovoltaic potential may not be as high as in other cities, due to the large proportion of less stable logistics buildings.

Figure 2.14. Time profile of self-production/consumption with solar panel (PV) and storage

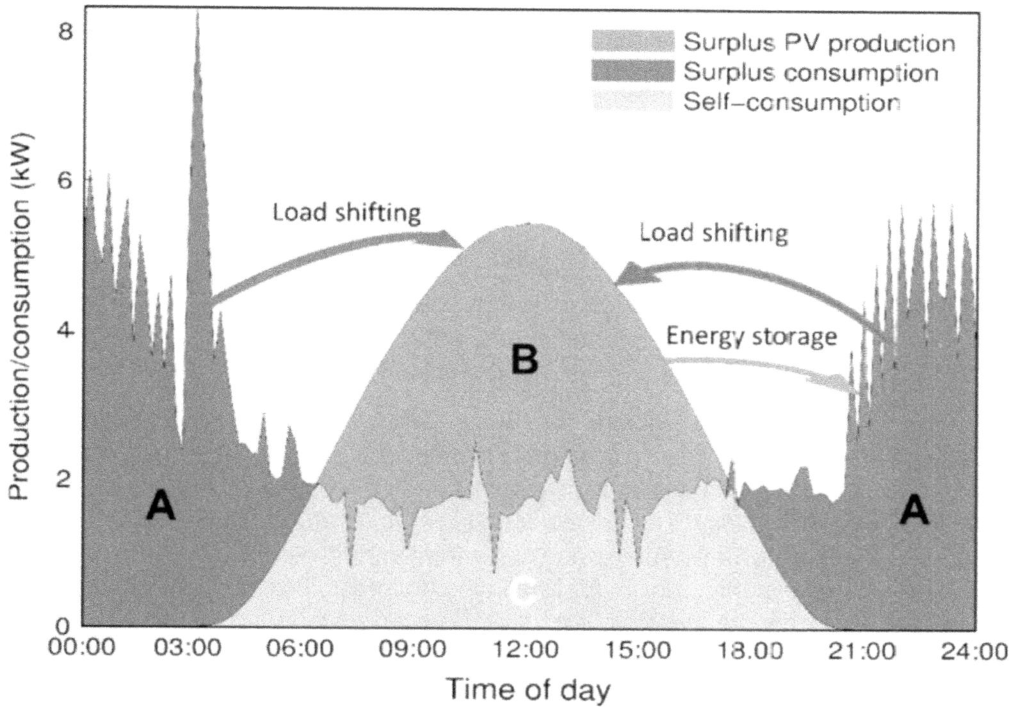

Source: (Alrawi et al., 2022[44])

In Hamburg, several obstacles impede rooftop solar panel installations, with a key challenge being limited access to information regarding technology, regulations, implementation and cost-saving potentials from rooftop installations, even today. For example, businesses considering rooftop solar panels could collaborate to collectively share the cost of connecting to the grid. Some companies in Hamburg are already working to assist companies to use self-produced VREs (Box 2.8). To enhance companies' awareness and information accessibility, it is key to actively promote and communicate the existence of support institutions that facilitate solar panel installations. These include the chambers of commerce, government-initiated initiatives such as the Hamburger Energielotsen, as well as private organizations like property owners' associations and coalitions like the Solar Offensive, which bring together various stakeholders. The HCC could play a key role in making the business case to its members. Including by pointing out to business models to share investment effort and benefits between building owners and tenants.

Box 2.8. Sager & Deus and Opländer Haustechnik supports businesses to integrate VRE into their energy mix

The Hamburg-based companies **Sager & Deus** and **Opländer Haustechnik** have made it their business to replace oil and gas-fired technologies. This includes maximum energy efficiency, the use of renewable energies and environmental technologies, as well as the offer of sanitary technologies that are economically justifiable and ecologically sustainable.

They currently employ 183 people and generate a total annual turnover of around 28 million euros. The companies' main focus is on the transformation of energy systems towards decentralised renewable systems. The goal is to generate renewable, emission-free electricity and heat as close to the consumers as possible. Electricity and heat consumers are becoming prosumers. The objective is to allow more market participants to benefit from value created in this way.

Sager & Deus focuses on renewable and citizen-oriented energy generation systems, serving individual houses, estates and larger commercial properties, as well as on the construction and operation of local collective heat pumps, fuel cells operating with hydrogen as well as combined heat and power plants.

Opländer works with heating, plumbing, ventilation, air conditioning, refrigeration and electric installations. In addition to building these systems, maintenance is an important part of their work. They deploy technologies serving the transition to climate neutrality such as combined heat and power, solar thermal and heat pump technology. The company endeavours to advance climate neutrality, helping to change mindset towards renewable energy and away from fossil fuels, including in craft businesses. It also endeavours to make crafts, which are important to reach climate neutrality, such as in the installation of renewable energy systems, attractive especially for young people. It aims to broaden participation in the environmental transitions, helping to shape a transition shaped more by the society at large and not only by central structures.

Storage can boost flexibility in electricity demand

Electricity storage is one of the most important elements to add the flexibility needed by VRE and sector coupling. Storing electricity is more costly and technically demanding than storing fossil fuels (Erbach, 2019[32]). The most relevant types of energy storage for energy-using companies are electrochemical batteries, hydrogen energy storage, and thermal energy storage (Ould Amrouche et al., 2016[45]) (Münster et al., 2020[46]).

Electrochemical batteries are highly heterogeneous and can be used on a wide range of scales. Batteries are especially well-suited for companies in modestly energy-intensive sectors, that generate their own electricity from solar rooftop systems or participate in demand-response programs. The most widely used types are lithium-ion (Li-ion) batteries, with a market share of 97% in 2017 in Germany. Li-ion batteries are also widely use in electric vehicles.

Battery prices fell by 98% between 1991 and 2018. Even the prices of small-scale residential battery systems fell by 70% in Germany from 2014 to 2020 (International Renewable Energy Agency, 2023[47]). Their prices especially for Li-ion batteries are expected to continue falling for several reasons, such as higher scale production capacity, better materials, or more competitive supply chains (Renewable Energy Agency, 2017[48]). However, environmental footprints of battery production are marked, which may push up prices when these are priced in, making demand response and sector coupling using installations serving also other purposes (heat reservoirs, vehicle batteries) more interesting.

HES converts electricity into hydrogen. This storage technology is of interest to companies requiring long-term energy storage (He et al. 2021[49]). As of now, hydrogen projects are not profitable. However, hydrogen-based energy storage has also become more attractive thanks to the reduced cost of electrolysers (International Renewable Energy Agency, 2023[47]).

As energy is stored in cold or hot fluids, thermal storage can be used with power-to-heat technology to heat or cool buildings (Van Nuffel et al., 2018[34]). In this case, buildings or warehouses are heated when electricity demand is low or supply high, thanks to a smart thermostat. When demand peaks, the building's heating is switched off. This type of storage is particularly suited to companies with high heating/cooling requirements (e.g., supermarkets) and industrial groups. Supermarkets, which need electricity to cool food using refrigeration compressors, are a good illustration of this phenomenon. When electricity is plentiful, cooling compressors can run at full capacity (Van Nuffel et al., 2018[34]).

Cooperation is key to develop sector coupling on a larger scale

Companies located in the same or in nearby buildings can work together to develop infrastructure and benefit from economies of scale. Scale economies can apply to VRE production and storage: bigger shared installations may reduce costs compared to individual use of smaller installations (Kost et al., 2021[29]). Firms can create partnerships to share the up-front cost of solar panels with storage, electric vehicle smart charging systems, or even electrolyser infrastructure. For example, Siemens has collaborated with the Hamburg Port Authority to use surplus electricity generated by wind turbines to produce hydrogen for heavy vehicles and ships.

Companies can engage in collaborative efforts by becoming participants in energy communities. Joining these communities has also enabled businesses to benefit from lower electricity costs, maximising the potential scale economies. These can materialise in shared self-production, including the bigger potential benefit from sector coupling resulting from varying electricity use profiles (e.g., commercial, and residential) of each member in the community. These communities are formed either around shared wind and solar generation setups or by establishing fully self-sustained microgrids. A noteworthy example is the community-oriented Virtual Power Plant in Ghent, which has effectively integrated over 100 PV systems with energy management systems and storage capabilities. Yet, businesses in Hamburg report that energy communities are difficult to make happen. Digital technologies, such as blockchain, have the potential to facilitate the formation of energy communities by guaranteeing secure peer-to-peer (P2P) and peer-to-utility transactions (OECD, 2021[50]).

Finally, coupling sectors requires different types of knowledge. Companies can share their expertise through partnerships. For example, Siemens has created a partnership with Evonik, a chemical company, to develop a technology for converting electricity into hydrogen. E.ON, an energy company, and Audi, a car manufacturer, have collaborated to develop a project combining residential solar photovoltaic systems, energy storage systems and electric vehicles.

Key actions

Immediate action:

- The HCC could provide guidance on the profitability of business on-site electricity production, especially solar PV rooftop, as well as on business models to take advantage of production potentials.
- Businesses should assess all potentials for own electricity production, especially solar PV rooftop.
- Businesses need to assess the potential for making their electricity use more flexible, taking into account investment in equipment they need for energy efficiency and electrification, including heat pumps and electric vehicles.
- Businesses can assess scope for cooperating with other economic agents in shared buildings or neighbourhoods in improving flexibility of energy use.

- The HCC can argue for market-based solutions for the regulation of electricity markets with increasing renewables shares, notably for granular pricing in space and time, in wholesale and retail markets.
- The HCC could conduct an inventory of demand response programmes and advocate the standardisation of processes and the provision of clear information on the regulations and different contracts that may exist.

By 2030

- The HCC should monitor solar rooftop electricity production in Hamburg as well as identify any barriers to expansion.
- The HCC could complete a review of demand response programmes, making sure they are implemented as much as possible, both at the individual business level, working with utilities, the regional and the national government.
- The HCC can provide guidance on all these steps as well as working groups to facilitate knowledge-sharing and coordination.

By 2040

- With renewables shares in German electricity production planned to rise to 80%, all businesses should realise their own production and consumption potentials.

Decarbonising buildings

The buildings sector has a key role to meet energy and climate policy goals, as in 2021 it accounted for around 49% of final energy consumption in Hamburg (26% for private households and 23% for non-industrial commercial buildings), and around 24% of Scope 1 greenhouse gas emissions (equally split between private households and commercial buildings). Private households and non-industrial commercial buildings accounted for 45% of Scope 2 emissions in Hamburg in 2021 (23% and 22% respectively), where most of these emissions are from heating and use of appliances in buildings (Hamburg Statistics Office, 2023[51]). A significant gap exists between ambition and actual emission reductions in the buildings sector: for several years, the sector has exceeded the emissions limits of the German Climate Protection Law.

Reaching climate neutrality in 2040 implies that oil and gas use need to be phased out. This is relevant for buildings used by businesses for commercial activity as well as for businesses letting housing to private households. In 2019, oil and gas accounted for more than 40% of total energy consumption in Hamburg's commercial buildings (12% and 31% respectively) (Hamburg Statistics Office, 2022[52]). From 2026, every newly installed heating system may be required by law to run on at least 65% renewable energy (Alkousaa and Kraemer, 2023[53]). This 65% regulation may not exclude gas boilers' installation from 2026 in existing buildings and in new buildings. However, given that boilers typically operate for 20-30 years (Müller and Langenheld, 2022[54]), new fossil fuel boilers are inconsistent with Hamburg's 2040 climate neutrality target and therefore create stranded asset risks. Delayed action will also lead to 'bunching' in the demand for installing needed equipment later and increased reequipment costs. Though heat pump and biomass heating system sales increased by 53% and 17% respectively in 2022 compared to 2021, and gas boiler sales decreased by 8%, two thirds of all newly installed heating systems sold in Germany in 2022 still run on gas (61%) and oil (5.8%) (Kyllmann, 2023[55]). Based on projections, heat pumps sales increased by 48% in 2023 compared to 2022, but gas and oil boiler sales also increased, by 39% and 91% respectively (Figure 2.15; (Agora Energiewende, 2024[56])).

Figure 2.15. Installing oil and gas boilers jeopardises climate neutrality targets

Sales of decentralised heating systems for individual apartments in Germany since 2012

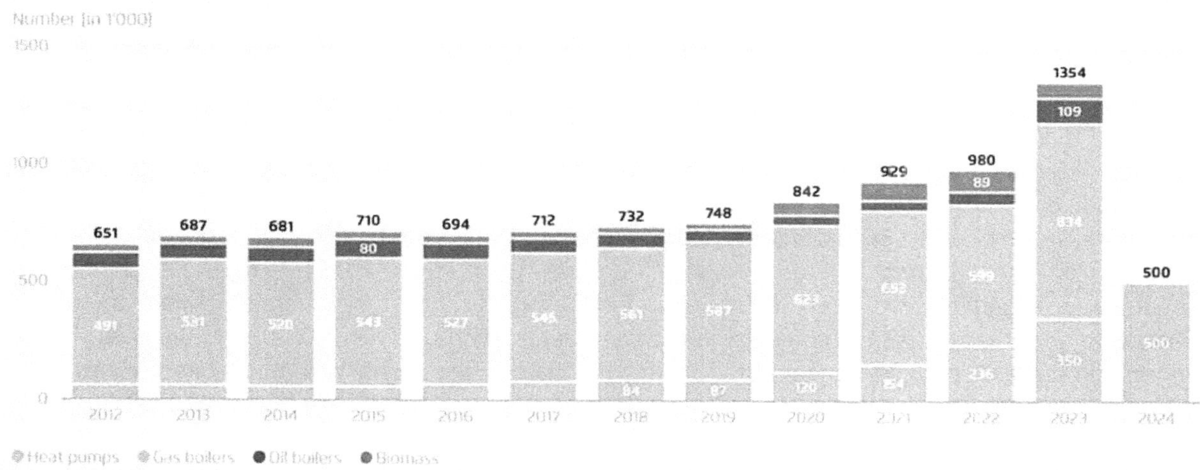

Note: Based on (Bundesverband der Deutschen Heizungsindustrie, 2023[57]). 2023: Projection, based on trends in Q1-Q3 of 2023. 2024: Climate-neutrality-compatible ramp-up path of heat pumps.
Source: (Agora Energiewende, 2024[56]).

To decarbonise buildings, businesses need to focus on two main things. First, replacing fossil oil- and gas-fired heating systems with heat pumps. Second, renovating building envelopes to reduce energy consumption, to be able to accommodate electrification and switching to renewables. Businesses should expect tightening of energy efficiency regulations for existing buildings to reduce energy use. As for oil and gas boilers, waiting for tighter regulations to be implemented would result in more 'bunching' and higher renovation costs. Moreover, these measures are intensive in regional employment, so their implementation needs to start quickly to avoid bottlenecks in construction capacity and skilled workers.

Businesses, residents and workers will benefit from the renovation and reequipment of buildings for higher energy efficiency. Firms will benefit through reduced heating bills and will be less vulnerable to energy price variations or rationing. The size benefits are not always known to businesses especially smaller ones. Residents and workers will benefit through more comfortable indoor temperatures and better air quality, resulting in health benefits (Climate Action Tracker, 2016[58]). Building users would also still need to adopt energy-saving behaviour to lower energy consumption further. As rising renewable shares require more spatial granularity in electricity pricing, lower regional electricity demand might one day also mean lower regional electricity prices.

How businesses can decarbonise buildings

Converting heat supply from fossil fuels to heat pumps

In view of Hamburg's 2040 climate neutrality target, businesses should replace fossil oil- and gas-fired heating systems with heat pumps. Heat pumps will allow meeting the 2040 target as well as the 2026 renewable energy requirement, and should become the main source of heat supply to reach climate neutrality where district heating is not available. For example, simply supplementing a conventional boiler with a solar thermal heating system is not sufficient for meeting the 65% renewable energy regulation (Müller and Langenheld, 2022[54]).

Heat pumps are a suitable option across a wide range of building types. The technological progress of recent years has significantly expanded the scope for their application: for example, heat pumps now work efficiently in buildings that have only been partially renovated (Müller and Langenheld, 2022[54]). Small renovation steps like replacing the windows or old radiators are often sufficient (Miara, 2021[59]). Hybrid heat pumps that work in conjunction with fossil-fuel-based heating systems have no economic or environmental advantage over regular heat pumps in most cases (Müller and Langenheld, 2022[54]).

Gas boilers may operate with or alongside sustainable biomass use or eventually with hydrogen. However, heat pumps have significant advantages over sustainable biomass- and hydrogen-based heating systems, which pose risks in terms of availability, costs and resilience. Given climate neutrality targets across OECD countries and beyond, demand for sustainable biomass and hydrogen will grow strongly, pushing up prices, as several difficult-to-decarbonise sectors will need to rely on these energy sources, such as air transport, shipping, road transport and industry (Material Economics, 2019[60]). Biomass growth also competes with essential land uses, notably for food production and biodiversity protection, and is vulnerable to shocks, including from extreme climate events such as fires or draught. Similarly, "green hydrogen" production may be concentrated in regions across the globe with the highest renewables potential and will be internationally tradable and therefore subject to shocks that can be transmitted internationally. Moreover, hydrogen produced with renewables may imply an efficiency loss of about a third. Thus, biomass and hydrogen should not be used on a large-scale for heating. Reaching climate neutrality in 2040 will require energy in buildings to be renewable at 100%, raising the economic risks of these energy carriers.

Geothermal heat pumps are a viable option, but they do not work for all buildings and have high setup costs as they require ground loops over large areas or drilling bore holes (Kavanaugh, Gilbreath and Kilpatrick, 1995[61]).

There are some potential barriers to the installation of heat pumps. First, the shortage of skilled workers for the installation. Second, local regulations that protect historical monuments (*Denkmalschutz*) or that do not allow to alter the city's 'landscape' (*Städtebauliche Erhaltungsverordnungen*) – since heat pumps are installed on the outside of buildings –, as well as regulations that do not allow to increase rents to cover the investment cost, according to businesses (*Soziale Erhaltungsverordnungen*). Local regulations that protect historical monuments are also a barrier to the installation of rooftop solar panels (see above). There is no data on the share of commercial buildings falling under *Denkmalschutz*, *Städtebauliche Erhaltungsverordnungen* and *Soziale Erhaltungsverordnungen*, but *Denkmalschutz* and *Städtebauliche Erhaltungsverordnungen* alone could cover around one third of commercial buildings. Furthermore, *Städtebauliche Erhaltungsverordnungen* are not always transparent (OECD, 2023[62]). Third, demand for heat pumps currently exceeds supply. Fourth, financing processes are lengthy, and building owners may not use federal subsidies to the extent possible ((Hamburger Behörde für Stadtentwicklung und Wohnen, 2022[63]); (Müller and Langenheld, 2022[54])). And fifth, building owners are reluctant to invest in heat pumps as Hamburg's current electricity supply would not be sufficient if close-to-all buildings would switch to heat pumps.

Converting heat supply from fossil fuels to district heating

In addition to heat pumps, district heating needs to become the other main source of heat supply. This includes increasing connections to district heating networks, expanding existing networks and building new networks (Hamburger Behörde für Stadtentwicklung und Wohnen, 2022[63]). With the expansion and construction of district heating networks, districts can decarbonise more easily and less expensively than at the individual building level (Hamburger Behörde für Stadtentwicklung und Wohnen, 2022[64]). District heating can be connected to different kinds of energy sources, including large heat pumps, solar thermal energy, geothermal energy or waste heat from industrial plants and data centres (Wehrmann, 2023[65]).

Nevertheless, not all buildings can be connected to a district heating source or it may be costly to do so. Centralised district heating solutions require more piping infrastructure for the more distant locations, and are thus expensive and require more time to get connected. Moreover, the contribution district heating can make to reach climate neutrality depends on whether district heating sources are zero-emission consistent and can be expected to last as industrial activities may close down. At present, the energy sources of district heating are not climate neutral. District heating systems currently based on fossil fuels should be converted to green energy sources (Müller and Langenheld, 2022[54]). To the extent district heating relies on industrial sources, they require fallback options if industrial activities drop out. Climate-neutral industrial processes may also decrease available industrial waste heat for use in district heating (Manz, Fleiter and Eichhammer, 2023[66]). Other potential barriers include the shortage of skilled workers and construction capacity, supply chain issues and long approval processes (Hamburger Behörde für Stadtentwicklung und Wohnen, 2022[63]).

Renovating building envelopes

Switching from fossil oil- and gas-fired heating systems to heat pumps will increase electricity demand. Businesses need to renovate building envelopes to reduce energy consumption and increase heat pumps' efficiency. Renovating buildings' envelopes mainly includes better insulating external walls, better insulating roofs and floors, and replacing old windows with more energy efficient ones (Hamburger Behörde für Stadtentwicklung und Wohnen, 2022[63]). Materials of current buildings can be recycled and reused for the renovation.

Buildings that consume too much energy to install a heat pump should be renovated first. Buildings that are or will be connected to district heating do not need to be renovated immediately as they are near carbon-neutral in energy use. Easier measures, such as replacing windows, could be implemented first. Businesses can take advantage of events like a change of tenant or needed renovations for other purposes to renovate building envelopes.

The HCC targets climate neutrality by 2040 – five years earlier than the city of Hamburg and Germany. Reaching climate neutrality in Hamburg's commercial buildings does not require all buildings to be renovated by 2040, provided they are reequipped in a carbon-neutral way. Even so, Hamburg's and Germany's climate neutrality targets for 2045 require major efforts to reduce energy demand. Assuming 50% of commercial buildings will be connected to district heating by 2045 (Bürgerschaft der Freien und Hansestadt Hamburg, 2023[67]), at least 2.5% of commercial buildings should be renovated each year, until 2045. Currently, the renovation rate is estimated to be 0.7-0.9% of commercial buildings per year in Germany (OECD, 2023[62]).The Hamburger Behörde für Stadtentwicklung und Wohnen (BSW) assumes a renovation rate of 1.6% in 2030, 1.7% in 2035 and 1.8% in 2045 for commercial buildings – taking into account the current shortage of skilled workers ((Hamburger Behörde für Stadtentwicklung und Wohnen, 2022[63]),(Hamburger Behörde für Stadtentwicklung und Wohnen, 2022[68])).

Potential barriers to the renovation of building envelopes include insufficient building capacity and the shortage of skilled workers in construction and planning, as well as insufficient supply and cost increases in building materials. Addressing skills shortages is a key priority to scale up renovation rates. Furthermore, building owners may not use federal subsidies to the extent possible and financing processes can be lengthy (Hamburger Behörde für Stadtentwicklung und Wohnen, 2022[63]). Local banks also mention that applying to subsidies of the regional development bank to pay for energy efficiency investment is too complex as they are broken into many different small schemes (OECD, 2023[62]). Subsidy schemes could be streamlined and information about them made more easily accessible and understandable for firms, especially SMEs.

Optimising the operation of technology in buildings

In addition to renovating buildings' envelopes, technology's operation in buildings can be optimised to further reduce energy consumption in new and existing buildings. This for example includes installing easy-to-understand energy efficiency displays on heating systems, digital control systems and regularly inspecting heating systems. The *Hamburger Behörde für Stadtentwicklung und Wohnen* (BSW) estimates that such measures could save an average of 10% of the final energy demand and that the investment pays off after ten years (Hamburger Behörde für Stadtentwicklung und Wohnen, 2022[63]). The lack of awareness of these technologies' potential can hold back implementation.

Quick low-cost options to be harnessed right away

Last, commercial building users should adopt the same energy-saving behaviours as in residential buildings, like lowering room temperatures, ventilating correctly, etc. In the case of lighting, switching to LEDs has the greatest savings potential (Hamburger Behörde für Stadtentwicklung und Wohnen, 2022[63]). These various measures could save approximately 5% of energy demand (Hamburger Behörde für Stadtentwicklung und Wohnen, 2022[63]). The main barriers are the lack of acceptance from users and lack of training opportunities for facility managers.

Governance of building decarbonisation: the Dutch example

Challenges and decarbonisation measures differ across buildings and neighbourhoods. For instance, as already mentioned, not all buildings can install a heat pump, or they may not be able to connect to district heating. To address energy efficiency renovation and reequipment in a targeted way, the Netherlands adopted a neighbourhood approach. Jointly launched by the national government, the Association of Dutch Municipalities as well as other bodies, the Natural Gas-Free Neighbourhood programme (PAW) provided up to EUR 5 million to each of 66 selected neighbourhoods across the Netherlands to help them experiment and scale up natural-gas-free measures (OECD, 2023[69]).

Neighbourhood-level communication has proven to be effective in the Netherlands, for example. Many neighbourhoods have promoted residents' success stories on heat pumps or insulation and have encouraged residents to form groups of 'neighbourhood ambassadors' to spread good practices. The municipality of Leusden hired a neighbourhood counsellor to oversee citizen initiatives and work as an intermediary between them and the municipality. Other municipalities have offered credible data to residents. Rotterdam, for example, developed a map to show the cheapest alternative solutions to natural gas based on where buildings are located (OECD, 2023[69]).

The HCC could help promote such initiatives.

Skills shortages and capacity constraints

The shortage of skilled workers in the construction sector and in renovation and heating professions poses a serious challenge to buildings' decarbonisation. The attractiveness of technical training, both vocational and academic, should thus be increased, for example with new entry-pathway programmes and qualifications in the fields of heat pumps, renewable energies and energy efficiency renovation.

Qualifications for construction jobs are currently rigid and broad in scope. For instance, photovoltaic installers need to be qualified for many tasks (such as complex calculations) beyond those needed for photovoltaic installation. This might slow down the needed expansion of qualifications, pushing up costs of investing in energy efficiency For example, it might lead them to drop out or deter them to apply (OECD, 2023[62]). Qualifications could be more flexible, for example allowing workers to choose between shorter trainings for specific tasks and longer trainings for construction work more generally. Workers opting for shorter training schemes should have the opportunity to do further training later on in their careers.

Moreover, barriers to the creation of a construction business should be removed (OECD, 2023[70]), and associated bureaucratic procedures digitised and simplified.

In the area of training, it is welcome that the crafts are planning to develop the new training qualification "Certified Professional Heat Pump Specialist". The new apprenticeship "Electronics Technician for Building System Integration" has already been implemented. With the "Heat Pump Development Programme", the government wants to promote the participation of trained specialists in the crafts as well as planners and energy consultants in further training courses specifically on the subject of heat pumps from 2023.

In practical terms, the industry is working on using synergies. This means that the central associations of crafts, sanitary, heating and air-conditioning technology, and electrical engineering have concluded an agreement to better design the installation of heat pumps across the crafts. This is intended to shorten installation times and make efficient use of scarce specialist capacities (Bundesverband der Deutschen Heizungsindustrie, 2023[71]).

To address skilled labour shortages, it is also crucial to support the labour market integration of women and upskilling of low-skilled workers, facilitate skilled labour immigration and expand adult learning opportunities (Table 2.5; Box 2.9; (OECD, 2023[70])). For example, the HCC is currently developing cooperation agreements with chambers of commerce in other countries, such as Uzbekistan, where construction workers start their training in their home country along with German language courses and then complete the training and start working in Germany. Better informing students early-on that 'green' construction occupations are in high demand is also key (OECD, 2023[70]).

Table 2.5. Past OECD recommendations and actions taken on training, education and labour market policies in Germany

Recommendations	Actions taken
Strengthen support for unskilled adults to obtain professional qualifications.	The 2023 reform of basic income support for jobseekers (introducing the new citizen's benefit *Buergergeld*) is a major step forward. It prioritises training over job uptake and improves financial support for longer-term training and education courses for job seekers to obtain professional qualifications. It also aims to improve access to basic education for jobseekers.
Provide financial incentives for employers to provide workplace learning for the low-skilled.	The government plans to significantly improve financial support for employees to participate in adult learning and to obtain professional degrees.
Facilitate participation of low-skilled individuals in adult education by taking further steps to validate uncertified skills, including those acquired-on-the job, and through workplace outreach.	The Federal Ministry of Education is planning to expand the pilot project ValiKom.
Improve transparency in the adult education market and facilitate access to guidance on adult training. Carefully monitor the outcome of financial support programmes for adult learning and education.	A National Continuing Education Online Platform ("Nationale Online-Weiterbildungsplattform - NOW") is being developed to increase transparency in and access to adult learning by providing appropriate information on courses, financing opportunities and skill needs in the labour market. The platform is planned to be launched in early 2024.
Liberalise occupational entry conditions, prioritising sectors subject to supply constraints (such as construction) and preserving the strengths of the vocational education and training system.	In 2020, the obligation to hold a master's craftsman degree when owning a craftsman business was reintroduced in 12 occupations, which restricted entry conditions further.

Source: (OECD, 2023[70]) (OECD, 2023[62]).

Box 2.9. Tackling skills shortages at the local level: examples from Denmark and Spain

The following initiatives, though not specifically targeted at the construction sector, could apply to it.

Strategy to close 'green skills gaps' in Copenhagen, Denmark

The City of Copenhagen collaborated with a range of local and regional stakeholders to develop a strategy to close 'green skills gaps'. The strategy supports several initiatives, including:

Initiatives to demonstrate 'green career' opportunities offered by vocational education institutions to attract young people; improve career guidance on 'green education' for youth; increase female participation in relevant professions; and support businesses in reskilling their employees.

Initiatives to develop 'green upskilling courses'; collaboration with firms to offer 'green internships' to unemployed people; guidance for unemployed people to steer them into green upskilling courses and internships.

A training centre with a focus on 'green skills' in Navarre, Spain

At the end of the 1990s, the Spanish region of Navarre started to invest heavily in renewable energy. Its electricity production from renewable sources expanded from zero in 1994 to 65% in 2009. This resulted in increased demand for renewable energy specialists, such as wind power maintenance staff. A lack of skilled workers became a barrier to the further expansion of the renewable energy industry.

In response, the regional government, in collaboration with the Navarre Confederation of Entrepreneurs and the Navarre Industry Association, set up CENIFER, a public training centre for renewable energies and energy efficiency. The Centre helped address skills shortages in the region by offering training adapted to businesses' needs that differed in length and included strong practical components. CENIFER now offers a wide range of courses for professions such as technicians in power plants, water management, thermal and fluid installations, solar thermal energy, etc.

Source: (OECD, 2023[72])

Financing the decarbonisation of buildings

Financial instruments to decarbonise buildings can include both traditional building financing instruments and mechanisms designed specifically for decarbonisation measures.

Traditional instruments include equity and self-financing, equipment leases, commercial debt and bonds. Nevertheless, the scale and pace at which the net-zero transition needs to happen requires these instruments to evolve to incentivise energy-efficient renovation and reequipment. 'Green' equivalents to these instruments can provide such incentives (LaSalle, 2022[73]). For example, 'green' or 'energy efficient' mortgage schemes offer discounted rates to better performing buildings or offer additional loans for energy efficiency improvements. Else, 'solar leases' allow businesses to install photovoltaic panels with no upfront cost while paying monthly rents for the panels. Depending on whether it is an operating or capital lease, building owners can buy the equipment at a reduced price at the end of the contract (LaSalle, 2022[73]).

Specialised instruments for buildings' decarbonisation harness the fact that energy-efficient buildings require less energy to operate. Typically, these instruments seek long-term profitability through energy efficiency improvements and on-site energy generation that reduce utility bills (LaSalle, 2022[73]). For example, in 'on-bill' financing and repayment schemes, utility companies or third-party lenders pay for the upgrades, and customers then repay the lender through their utility bills. Else, Property Assessed Clean

Energy (PACE) programmes allow building owners to pay the costs back over time through property tax bills (LaSalle, 2022[73]).

New business models that reduce the upfront capital cost associated with the most energy-efficient and low-carbon buildings technologies are critical (IEA, 2023[74]). The HCC could explore further financing schemes for the decarbonisation of commercial buildings.

Grants and support programmes will have to be adapted to make sure all businesses, in particular SMEs, can afford replacing their heating systems to meet new regulations. Policies should support specifically the installation of heat pumps in buildings with historically low rates of installation, such as apartment buildings and buildings with communal heating systems (Müller and Langenheld, 2022[54]).

Raising building heights increases real estate values, which can also be used to finance needed energy-saving investment through land-based finance instruments. Land-based finance enables local governments to recover land value increases resulting from public infrastructure provision and changes in land-use regulations (OECD, 2022[75]).

Key actions

Immediate actions

- Businesses should assess all fossil-fuel fired heating systems 20 years or older, requiring replacement in the near term, and identify needed actions to replace them with heat pumps or district heating, including needed energy efficiency investment.
- Businesses should replace heating systems that depend on fossil fuels with renewable energy systems, notably heat pumps, including in existing buildings, while avoiding installations that will depend on hydrogen or biomass firing.
- Businesses should aim to have a renovation rate of 2.5% of buildings per year by 2030.
- The HCC could help identify buildings that need to be renovated first to improve energy efficiency, such as buildings with high energy consumption, without access to district heating or that are easier to renovate. Businesses can take advantage of events like a change of tenant or needed renovations for other purposes to renovate building envelopes.
- The HCC could help identify bottlenecks in skills and construction capacity as well as policies to address them.
- The HCC could help assess financing gaps and suggest new funding instruments, especially for SMEs.
- The HCC could help assess barriers to the installation of heat pumps and solar panels in building regulations.
- The HCC could raise awareness and provide guidance on all these steps as well as create working groups and networks to facilitate knowledge-sharing and coordination among firms.
- Public companies could take a leading role to show the example and serve as role models.

2030

- The HCC could help assess whether all steps are taken to ensure skills gaps, construction capacity constraints and financing gaps are addressed.

2040

- Businesses should complete replacement of all fossil-fuel based heating systems in commercial buildings and commercially let apartments with renewable alternatives.

References

Agora Energiewende (2024), *Die Energiewende in Deutschland: Stand der Dinge 2023. Rückblick auf die wesentlichen Entwicklungen sowie Ausblick auf 2024*, https://www.agora-energiewende.de/fileadmin/Projekte/2023/2023-35_DE_JAW23/A-EW_317_JAW23_WEB.pdf. [56]

Agora Energiewende (2023), , https://static.agora-energiewende.de/fileadmin/Projekte/2022/2022-04_DE_Scaling_up_heat_pumps/2022_Scaling_up_heat_pumps_in_Germany.pdf. [81]

Airbus (2022), *Progressing with purpose: Annual report*. [10]

Albadi, M. and E. El-Saadany (2008), *A summary of demand response in electricity markets*, https://doi.org/10.1016/j.epsr.2008.04.002. [39]

Alkousaa, R. and C. Kraemer (2023), *German parliament passes law to phase out gas and oil heating*, https://www.reuters.com/world/europe/german-parliament-passes-watered-down-heating-law-2023-09-08/#:~:text=Some%20399%20parliamentarians%20voted%20in,run%20on%2065%25%20renewable%20energy. [53]

Alrawi, O. et al. (2022), "Economic Viability of Rooftop Photovoltaic Systems and Energy Storage Systems in Qatar", *Energies*, Vol. 15/9, https://doi.org/10.3390/en15093040. [44]

ArcelorMittal (2021), *Climate Action Report*. [11]

Aurubis (2023), *Enabling tomorrow: sustainability report*. [4]

Beiersdorf (2022), *Formula for sustainable progress*. [9]

Bundesverband der Deutschen Heizungsindustrie (2023), *Heizungsmarkt weiter auf Rekordniveau*, https://www.bdh-industrie.de/presse/pressemeldungen/artikel/heizungsmarkt-weiter-auf-rekordniveau. [57]

Bundesverband der Deutschen Heizungsindustrie (2023), *Jahresbilanz 2022: Heizungsmarkt boomt*, https://www.bdh-industrie.de/fileadmin/user_upload/Pressemeldungen/PM_BDH_Jahresbilanz_Heizungsindustrie_13022023b.pdf. [71]

Bürgerschaft der Freien und Hansestadt Hamburg (2023), *Mitteilung des Senats an die Bürgerschaft: Zweite Fortschreibung des Hamburger Klimaplans*, https://www.hamburg.de/contentblob/17316156/81785d14d09a5942af1d6c0fccf59c9e/data/d-zweite-fortschreibung-hamburger-klimaplan.pdf. [67]

Climate Action Tracker (2016), *Constructing the Future: Will the Building Sector Use its Decarbonisation Tools?*, https://climateanalytics.org/publications/2016/constructing-the-future-will-the-building-sector-use-its-decarbonisation-tools/. [58]

ENEFIRST (2020), *Report on international experiences with E1st. Deliverable D2.2 of the ENEFIRST project, funded by the H2020 programme.*, http://enefirst.eu/ (accessed on 13 October 2023). [40]

Energy Agency, I. (n.d.), *Unlocking the Potential of Distributed Energy Resources Power system opportunities and best practices*, http://www.iea.org/t&c/. [80]

Erbach, G. (2019), *Energy storage and sector coupling: Towards an integrated, decarbonised energy system*, European Parliamentary Research Service. [32]

European Commission (2022), *Carbon Border Adjustment Mechanism*. [18]

European Commission (2022), *Green Deal: EU agrees law to fight global deforestation and forest degradation driven by EU production and consumption*. [16]

European Commission (2022), *Just and sustainable economy: Commission lays down rules for companies to respect human rights and environment in global value chains*. [14]

Gea-Bermúdez, J. et al. (2021), “The role of sector coupling in the green transition: A least-cost energy system development in Northern-central Europe towards 2050”, *Applied Energy*, Vol. 289, https://doi.org/10.1016/j.apenergy.2021.116685. [33]

Greenhouse Gas Protocol (2023), *Life Cycle Databases*. [21]

Greenhouse Gas Protocol (2013), *Technical Guidance for Calculation Scope 3 Emissions*. [20]

Hamburg Klimabeirat (2023), *Photovoltaik in Hamburg*. [27]

Hamburg Statistics Office (2023), , https://www.statistik-nord.de/zahlen-fakten/umwelt-energie/energie/dokumentenansicht/product/3381/energie-und-co2-bilanzen-fuer-hamburg-361?cHash=4201529a752424c94a05eb3c4ae751ea. [51]

Hamburg Statistics Office (2022), . [52]

Hamburger Behörde für Stadtentwicklung und Wohnen (2022), *Steckbrief Gewerbe, Handel und Dienstleistungen sowie übrige Verbraucher*. [63]

Hamburger Behörde für Stadtentwicklung und Wohnen (2022), *Steckbrief Sektor Private Haushalte*. [68]

Hamburger Behörde für Stadtentwicklung und Wohnen (2022), *Steckbrief Sektorübergreifend*. [64]

Hapag-Lloyd (2022), *Making a difference in the future of shipping*. [3]

Hargroves, K. et al. (2023), “The Role of Distributed Energy Resources and Associated Business Models in the Decentralised Energy Transition: A Review”, *Energies*, Vol. 16/10, https://doi.org/10.3390/en16104231. [41]

He, G. et al. (2021), “Sector coupling: Via hydrogen to lower the cost of energy system decarbonization”, *Energy and Environmental Science*, Vol. 14/9, pp. 4635-4646, https://doi.org/10.1039/d1ee00627d. [49]

Hossain, M. et al. (2016), *Role of smart grid in renewable energy: An overview*, Elsevier Ltd, https://doi.org/10.1016/j.rser.2015.09.098. [30]

IEA (2023), *Energy system - Buildings*, https://www.iea.org/energy-system/buildings. [74]

IEA (2023), *How can smart charging steer electric vehicle uptake in India? – Analysis - IEA*, https://www.iea.org/commentaries/how-can-smart-charging-steer-electric-vehicle-uptake-in-india (accessed on 18 July 2023). [36]

IEA (2022), *Renewables 2022: Analysis and forecast to 2027*, OECD Publishing, Paris, https://doi.org/10.1787/96bc279a-en. [43]

IEA (2022), *Steering Electricity Markets Towards a Rapid Decarbonisation*, OECD Publishing, Paris, https://doi.org/10.1787/d2f0068c-en. [38]

IEA (2022), *World Energy Outlook 2022*, OECD Publishing, Paris, https://doi.org/10.1787/3a469970-en. [26]

IEA (2022), *World Energy Outlook Special Report The Future of Heat Pumps*, http://www.iea.org. [78]

International Energy Agency (2022), *Raw materials strategy of the Federal Government: Securing a sustainable supply of non-energy mineral raw materials for Germany*. [17]

International Renewable Energy Agency, T. (2023), *LOW-COST FINANCE FOR THE ENERGY TRANSITION*. [47]

Kavanaugh, S., C. Gilbreath and J. Kilpatrick (1995), *Cost Containment for Ground-source heat pumps*, https://www.osti.gov/etdeweb/servlets/purl/895130. [61]

Kost, C. et al. (2021), *LEVELIZED COST OF ELECTRICITY RENEWABLE ENERGY TECHNOLOGIES*, http://www.ise.fraunhofer.de. [29]

Kraftnät, S. (2017), *Finer time resolution in Nordic power markets: A Cost Benefit Analysis*. [37]

Krane, J. and R. Idel (2021), *More transitions, less risk: How renewable energy reduces risks from mining, trade and political dependence*, Elsevier Ltd, https://doi.org/10.1016/j.erss.2021.102311. [35]

Kyllmann, C. (2023), *Clean Energy Wire*, https://www.cleanenergywire.org/news/gas-boilers-continue-dominate-german-heating-market-heat-pumps-rise. [55]

LaSalle, J. (2022), *Financing Net Zero Carbon Buildings*, https://www.climatepolicyinitiative.org/wp-content/uploads/2022/07/Buildings-Scoping-Paper-final.pdf. [73]

Manz, P., T. Fleiter and W. Eichhammer (2023), "The effect of low-carbon processes on industrial excess heat potentials for district heating in the EU: A GIS-based analysis", *Smart Energy*, Vol. 10, https://doi.org/10.1016/j.segy.2023.100103. [66]

Marquad & Bahls (2020), *Sustainability Report*. [8]

Material Economics (2019), *Industrial Transformation 2050 - Pathways to Net-Zero Emissions*, https://materialeconomics.com/publications/industrialtransformation-2050. [60]

Maxingvest AG (2022), *Maxingvest AG*. [7]

Miara, M. (2021), *Innovation4E*, https://blog.innovation4e.de/en/2021/02/24/does-a-house-have-to-be-renovated-first-in-order-to-install-a-heat-pump/. [59]

Müller, S. and A. Langenheld (2022), *Agora Energiewende*, https://www.agora-energiewende.de/en/blog/a-heat-pump-revolution-for-germany-1/. [54]

Münster, M. et al. (2020), *Sector Coupling: Concepts, State-of-the-art and Perspectives*. [46]

NEA/IEA (2020), *Projected Costs of Generating Electricity 2020*, OECD Publishing, Paris, https://doi.org/10.1787/a6002f3b-en. [28]

Nordex (2022), *Together for change - wind for a sustainable future.* [12]

OECD (2023), "Decarbonising homes in cities in the Netherlands: A neighbourhood approach", *OECD Regional Development Papers* 42, https://doi.org/10.1787/b94727de-en. [69]

OECD (2023), *Job Creation and Local Economic Development 2023: Bridging the Great Green Divide*, https://doi.org/10.1787/21db61c1-en. [72]

OECD (2023), *OECD Economic Surveys: Germany 2023*, https://doi.org/10.1787/9642a3f5-en. [70]

OECD (2023), *OECD Environmental Performance Reviews: Germany 2023*, OECD Environmental Performance Reviews, OECD Publishing, Paris, https://doi.org/10.1787/f26da7da-en. [25]

OECD (2023), *OECD interviews with local stakeholders.* [62]

OECD (2023), *OECD SME and Entrepreneurship Outlook 2023*, OECD Publishing, Paris, https://doi.org/10.1787/342b8564-en. [24]

OECD (2022), "Financing SMEs for sustainability: Drivers, Constraints and Policies", *OECD SME and Entrepreneurship Papers*, No. 35, OECD Publishing, Paris, https://doi.org/10.1787/a5e94d92-en. [22]

OECD (2022), *Global Compendium of Land Value Capture Policies*, https://www.oecd-ilibrary.org/urban-rural-and-regional-development/global-compendium-of-land-value-capture-policies_4f9559ee-en. [75]

OECD (2021), *OECD instruments on responsible business conduct: supporting implementation and accountability of business' net-zero commitments.* [2]

OECD (2021), *OECD Regional Outlook 2021: Addressing COVID-19 and Moving to Net Zero Greenhouse Gas Emissions*, OECD Publishing, Paris, https://doi.org/10.1787/17017efe-en. [50]

OECD (2021), *Policy Trends in Environmental Due Diligence.* [13]

OECD (2019), *OECD Territorial Reviews: Hamburg Metropolitan Region, Germany*, OECD Territorial Reviews, OECD Publishing, Paris, https://doi.org/10.1787/29afa27f-en. [23]

Otto Group (2022), *Annual Report.* [5]

Ould Amrouche, S. et al. (2016), "Overview of energy storage in renewable energy systems", *International Journal of Hydrogen Energy*, Vol. 41/45, pp. 20914-20927, https://doi.org/10.1016/j.ijhydene.2016.06.243. [45]

Renewable Energy Agency, I. (2021), *Sector coupling in facilitating integration of variable renewable energy in cities sector coupling in facilitating integration of variable renewable energy in cities*, http://www.irena.org. [42]

Renewable Energy Agency, I. (2017), *ELECTRICITY STORAGE AND RENEWABLES: COSTS AND MARKETS TO 2030*, http://www.irena.org. [48]

Rinaldi, G. (2023), *Thermal turnaround: German government settles dispute over heating law*, https://www.politico.eu/article/germany-government-settle-dispute-heating-law/. [79]

Sedex (2023), *Germany's new Supply Chain Due Diligence Act: What you need to know*. [15]

Task Force on Climate-Related Financial Disclosures (2017), *Recommendations of the Task Force on Climate-related Financial Disclosures*. [76]

TCFD (2017), *Recommendations | Task Force on Climate-Related Financial Disclosures*, https://www.fsb-tcfd.org/recommendations/ (accessed on 19 July 2023). [1]

United Nations' High-Level Expert Group (2022), *Integrity Matters: Net Zero Commitments by Businesses, Financial Instituions, Cities and Regions*. [77]

Van Nuffel, L. et al. (2018), *Sector coupling: how can it be enhanced in the EU to foster grid stability and decarbonise?*. [34]

Vanholme, S. et al. (2022), *Sector coupling: A key concept for accelerating the energy transformation*. [31]

Vattenfall (2022), *Leading the way to fossil freedom: Annual and sustainability report*. [6]

Wehrmann, B. (2023), *Clean Energy Wire*, https://www.cleanenergywire.org/news/about-half-german-households-could-use-district-heating-decarbonise-industry. [65]

World Economic Forum, OECD, Business at OECD (2023), *Emissions Measurement in Supply Chains: Business Realities and Challenges. White paper.*. [19]

Note

[1] SMEs are companies with fewer than 250 employees.

Annex 2.A. Questionnaire for Hamburg Businesses

General information

1. Which sector does your company belong to?

- Industry
- Construction
- Trade
- Hotel and catering
- Logistics and transport
- Financial sector
- IT and media
- Healthcare
- Other services
- Public administration

2. How many employees are working in your company?

- 0 to 3 employees
- 4 to 9 employees
- 10 to 19 employees
- 20 to 49 employees
- 50 to 99 employees
- 100 to 249 employees
- 250 to 499 employees
- 500 and more employees

Climate protection activities of your company

3. How important is climate protection for your business?

□ Not important at all □ Little important □ Somewhat important □ Important □ Very important

4. Has your company set a net zero GHG emissions target? □ Yes □ No

If yes:

5.a. By when:

□ 2040 or earlier □2041-2050 □ Later □ Don't know

5.b. Has your company announced its net zero target publicly?

□ Yes, already published □ No, but planning to do so □ No □ Don't know

5.c. Has your company set intermediate targets?

□ Yes □ No

5.d. For which scope of emissions have you set targets?

□ Scope 1 – direct GHG emissions from the operations of the company

□ Scope 2 – indirect GHG emissions from the consumption of electricity, heat, and steam of the company

□ Scope 3 – all other indirect emissions in the value chain (upstream and downstream) of the company

□ Don't know

5.e. Does your company have an action plan to achieve climate neutrality?

□ In implementation

□ In planning

□ No action plan

5.f. What measures does your company plan to undertake to achieve climate neutrality?

(Multiple answers possible)

Define responsibilities in management

□ Yes, already implemented

□ Yes, in implementation or planning

□ No

Include the climate neutrality target in human resource planning

□ Yes, already implemented

□ Yes, in implementation or planning

□ No

Invest in climate-neutral technologies, including renewable energies and their flexible use when wind and solar energy are available

□ Yes, already implemented

□ Yes, in implementation or planning

□ No

Reduction of energy consumption

□ Yes, already implemented

□ Yes, in implementation or planning

□ No

Reduction of material use, circular economy

□ Yes, already implemented

□ Yes, in implementation or planning

□ No

New, climate-neutral products in sales

□ Yes, already implemented

□ Yes, in implementation or planning

□ No

Climate-neutral products in purchasing

□ Yes, already implemented

□ Yes, in implementation or planning

□ No

Other

□ Yes, already implemented

□ Yes, in implementation or planning

□ No

If no [has not set a net-zero target]:

5.g. Why has your company not set a net zero GHG emission target?

(Multiple answers possible)

□ In development

□ Lack of resources to identify a target (in need of more support)

□ No benefit for the company

□ Leads to competitive disadvantages

□ Not relevant for the company

□ Don't know

□ Other

5.h. Is financing greenhouse gas emission reductions elsewhere (such as financing reforestation activities or greenhouse gas emission reductions of other companies) to be a major part (one third or more) of the emission reduction for achieving the climate neutrality target in your company?

□ Yes

□ No

□ Don't know

6. Has your company set explicit targets for phasing of fossil fuels and increasing the use of renewable energies?

□ Yes, phase out fossil fuels

□ Yes, expand renewable energies

□ Yes, both

□ No, neither

7. What business opportunities and challenges do you see for your company in the field of climate neutrality?

(Multiple answers possible)

Opportunities:

□ Better financing for investments in climate-neutral business models

□ Profit increase and/or competitive advantage through low-CO2 products

□ New and more efficient production processes

□ New business areas opened up

□ Employer attractiveness

□ Better product marketing

□ New suppliers, customers and networks

□ Other

□ No opportunities

Challenges:

□ Investment in climate-neutral production and its financing.

□ Uncertain impact on costs, sales and operations

□ Higher production costs

□ Lack of qualified labour and skilled workers

□ Restructuring of corporate structures and management

□ Threats and elimination of the business field

□ Loss of suppliers, customers and networks

□ Other

□ No challenges

8. Does your company monitor climate indicators (e.g. emissions, energy consumption) and disclose them?

□ Yes

□ No

□ Don't know

If yes:

Which climate indicators does your company monitor?

GHG emissions

Greenhouse gas emissions Scope 1 (direct greenhouse gas emissions from the company's business activities)

□ Monitoring □ Disclosing □ Neither monitoring nor disclosing

Greenhouse gas emissions Scope 2 (indirect greenhouse gas emissions from the consumption of electricity, heat and steam by the company)

□ Monitoring □ Disclosing □ Neither monitoring nor disclosing

Greenhouse gas emissions Scope 3 (all other indirect emissions in the value chain, upstream and downstream, of the company)

□ Monitoring □ Disclosing □ Neither monitoring nor disclosing

Emission reduction

□ Monitoring □ Disclosing □ Neither monitoring nor disclosing

Total energy consumption

□ Monitoring □ Disclosing □ Neither monitoring nor disclosing

Energy effiency

□ Monitoring □ Disclosing □ Neither monitoring nor disclosing

Consumption/production of renewable energy

□ Monitoring □ Disclosing □ Neither monitoring nor disclosing

Impact on ecosystems in supply chains

□ Monitoring □ Disclosing □ Neither monitoring nor disclosing

Revenues from low carbon products

□ Monitoring □ Disclosing □ Neither monitoring nor disclosing

Other

9. How would you evaluate the availability of the following support services on the way to climate neutrality?

Advisory bodies

□ Not available □ Available but insufficient □ Sufficiently available

Networks with businesses in the own value chain, such as suppliers or customers

□ Not available □ Available but insufficient □ Sufficiently available

Other networks such as with other businesses, with experts or research institutions

□ Not available □ Available but insufficient □ Sufficiently available

Financial support

□ Not available □ Available but insufficient □ Sufficiently available

10. Has your company adopted any of these circular business models?

□ **Replace traditional inputs with secondary material** (e.g. using recycled plastics for office furniture)

□ **Shared use of capital goods** including paying for a service instead of product ownership (e.g. printing services) or rent (e.g. company car)

□ **Refurbishment and remanufacturing** restoration of degraded products, either for a fee, or for subsequent resale to original or new owner

□ **Other**

□ **No Introduction of Circular Business Models**

11. What are the main obstacles hindering the implementation of circular business models in your company? (Select the 3 main obstacles)

□ Lack of technological solutions / opportunities

□ Lack of financial resources

□ Cultural barriers / Problems of acceptance

□ Lack of human resources

□ Lack of skills

□ Lack of a conducive regulatory framework

□ Lack of awareness of circular business models / Low receptiveness

□ Transaction costs to move from a linear ownership model to a more circular model

□ Lack of critical scale

If your company has collected experiences on the climate neutrality way (for example, your action plans or sustainability reports) that you would like to share, we would be happy to receive them. Information on circular economy business models that support this path are also welcome. Thank you very much!

Annex Table 2.A.1. Survey participation

Comparison of characteristics of businesses responding to the OECD-HCC Survey with the characteristics of HCC member businesses.

1.Comparison of business main sector of activity				
	Respondents	**Share of all respondents %**	**Number of HCC member companies**	**Share of all HCC member businesses %**
Industry	9	7.2%	5006	2.9%
Construction	10	8.0%	9002	5.1%
Trade	14	11.2%	32474	18.6%
Hopsitality	3	2.4%	6647	3.8%
Logistics and transport	15	12.0%	11553	6.6%
Financial sector	5	4.0%	14362	8.2%
IT and media	16	12.8%	15075	8.6%
Healthcare	8	6.4%	3164	1.8%
Other services	45	36.0%	77545	44.3%
Public administration	0	0.0%	204	0.1%
Total	125	100.0%	175032	100.0%
2. Comparison of employee size				
	Respondents	**Share of all respondents %**	**Number of HCC member companies**	**Share of all HCC member businesses %**
0 to 3	9	7.2%	157163	87.1%
4 to 9	8	6.4%	13959	7.7%
10 to 19	18	14.4%	3976	2.2%
20 to 49	25	20.0%	2886	1.6%
50 to 99	19	15.2%	1115	0.6%
100 to 499	18	14.4%	1004	0.6%
500 and more	28	22.4%	258	0.1%
Total	125	100.0%	180361	100.0%

Source: HCC

Annex Figure 2.A.1. Most businesses have not set net zero targets because of lack of resources

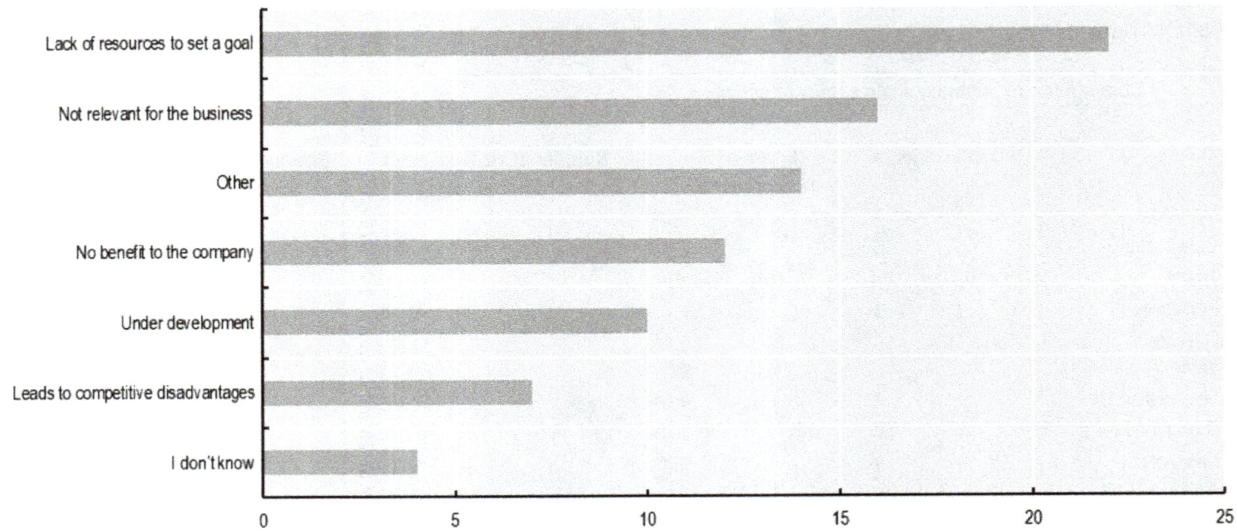

Source: Survey carried out by HCC (2023).

Annex Figure 2.A.2. Out of the businesses that have net zero targets, the majority have a climate neutrality goal of 2040 or before

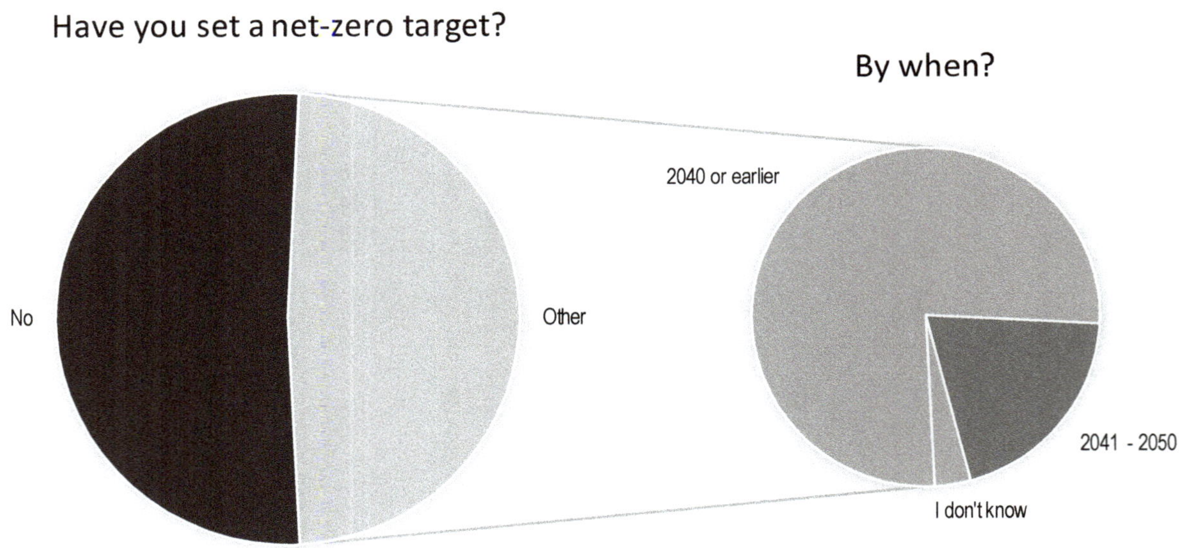

Source: Survey carried out by HCC (2023).

3 Reaching climate neutrality in freight and industry

The port of Hamburg has taken a leading role in reaching climate neutrality by 2040. This chapter shows that Hamburg's strategic location can turn it into a major climate-neutral transport centre, building on its strong rail infrastructure. In this way Hamburg can deliver climate-neutral transport services client businesses need throughout Western and Central Europe quickly and at low cost. This includes harnessing its potential for a hydrogen hub. The transition to zero-emission fuels for international shipping therefore needs to be prepared now. This will raise costs by 2030 on account of initially higher fuel production cost and needed investment in zero-emission shipping, but an energy-saving fuel mix may help keep longer-term costs and trade flow impacts modest. Maximising the use of rail would deliver benefits beyond Hamburg. It requires addressing railway infrastructure bottlenecks and digitalization. Quick major steps towards electrified road transport are also needed, with electric charging infrastructure and continued efforts to improve logistics. Hamburg can play an important role in delivering imported green hydrogen and hydrogen-derived products, via pipelines or ships, to serve demand in local industry and shipping as well as in neighbouring industrial regions.

Freight and logistics through ports play a critical role in trade flows. Goods handled in ports account for 46% of the value of goods traded between the EU and the rest of the world, and 75% of its volume (ERA, 2022[1]). Hamburg hosts one of the four largest ports in Europe. Transport and logistics activity in and around the port reflect its importance as a major economic hub. Hamburg is home to a large number of industrial, wholesale and retail businesses, including in e-commerce, which depend on the port and its transport and logistics services. Preparing the Hamburg port for climate-neutral international shipping is therefore central to the Hamburg economy, as well as beyond, providing much-needed zero-carbon transport services to business on the continent. The first section of this chapter is devoted to this issue.

Land-based transport connections from and to the port are of equal strategic importance. The vast cargo turnover in Hamburg has fostered a high-frequency supply chain in the hinterland. The influence of Hamburg and its port reaches deep into this hinterland down transport chains, shaping economic and environmental performance in the Hamburg region, Germany and many parts of Europe.

The major ports in Europe often serve overlapping hinterlands: as the distance between the port and the hinterland grows, competition between ports and between transport modes increases. Yet, land-based freight services are, alongside maritime shipping, particularly difficult to decarbonise. The second section therefore discusses the implications of Hamburg hinterland transport and logistics, providing insights into actions to successfully manage the transition to climate neutrality while harnessing Hamburg's comparative advantage and its strategic position as one of Europe's major ports.

The port is closely integrated with the manufacturing of basic materials that are also particularly challenging to make climate-neutral. This applies in particular to basic metals production and oil refining. These sectors depend strongly on fossil fuels, both as energy carriers and raw materials. They often require high temperatures in production processes which do not easily lend themselves to electrification. The decarbonisation of key manufacturing sectors in Hamburg is discussed in the third section.

Hydrogen and hydrogen-derived products are important for the climate-neutral operation of international maritime shipping as well as for some of Hamburg's manufacturing activities. Some manufacturing might also require carbon capture and storage (CCS). The Hamburg port may also serve as a hub for the provision of hydrogen, CCS and related transport services. This is discussed in the fourth section of the chapter.

Anticipating the impact of decarbonising maritime transport

The Hamburg port has already adopted a strategy to reach climate neutrality for Scope 1 and 2 emissions in port operations by 2040. This includes all emissions in freight handling and transport. A consistent CO2 balance for the port will be used to monitor and control this process (Behörde für Wirtschaft und Innovation, 2023[2]). The first part of this chapter will therefore focus on the decarbonisation of international shipping. Preparing the Hamburg port for the major transformations this will bring is key for the Hamburg economy. This section will deploy transport and macroeconomic modelling for the port of Hamburg to provide insights about the implications, in terms of zero-emission fuel supply and impacts on trade (Halim et al., 2018[3]).

In July 2023 the International Maritime Organization (IMO, Box 3.1) adopted the revised 2023 IMO GHG strategy. This strategy sets objectives to reach net-zero GHG emissions by or around 2050 and reduce GHG emissions by at least 20% by 2030 while striving to reduce them by 30% and reduce GHG emissions by 70% by 2040, compared to 2008.

According to the revised 2023 IMO GHG Strategy, global shipping must lower its carbon intensity (CO2 emissions per transport service) by at least 40% by 2030.

Box 3.1. The role of the International Maritime Organization in decarbonising maritime freight

The International Maritime Organization (IMO) is a UN agency with 174 member countries. It is an intergovernmental consultative body which adopts policy measures to prevent and reduce environmental impacts as well as to improve the safety and security of international shipping. IMO has stated its aim to align with the 2015 Paris Agreement. It has revised the 2018 "Strategy on reduction of GHG emissions from ships" in 2023. A further review will be finalised when the IMO's Marine Environment Protection Committee (MEPC) meets in the autumn of 2028 to adopt the 2028 IMO Strategy. The MEPC handles environmental concerns such as the management and prevention of ship-sourced pollution, or ship recycling. As a technical body to create, review, and revise the IMO strategy, the Intersessional Working Group on GHG Emissions (ISWG GHG) proposes policy measures to the MEPC.

Source: (IMO, 2023[4])

Economic measures to decarbonise international shipping are under ongoing discussion. To meet the agreed targets, IMO member countries have proposed baskets of economic and technical measures that will need to be adopted by 2025. These measures will have the broadest coverage, in terms of the number of countries as well as the range of ship types regulated under IMO policy. This section takes the latest IMO agreement as the starting point of the analysis, as well as the proposed basket of measures from the European Commission (EC).

For most businesses involved with the trade of goods from and to the port, the emissions from international shipping are Scope 3 emissions. Climate neutrality for international shipping by 2050 may therefore be consistent with the 2040 HCC climate neutrality target, provided there is no strong reliance on international offsets to reach it. However, businesses operating vessels in international shipping headquartered in Hamburg should move ships to largely zero-emissions fuels by 2040.

Preparing the port for zero-carbon fuels

The availability and the use of zero-emission fuels will be key to decarbonising international shipping according to the timeline that the revised IMO targets. Ports could play a strategic role as a logistics node in their supply, serving as bunkering stations for ships.

To estimate cost-minimising zero-carbon fuels, this section draws on the results of a modelling exercise carried out with the NavigaTE model (Mærsk Mc-Kinney Møller Center, 2021[5]). NavigaTe is a techno-economic shipping model developed by the Mærsk Mc-Kinney Møller Center for Zero Carbon Shipping to assess the most cost-effective decarbonisation pathways, taking comprehensive account of costs from adopting different fuel measures. It captures the entire energy value chain for powering the maritime vessel from raw materials and primary energy used in fuel production to the fuel combustion in the vessel. A description of this model can be found in the Annex. The model uses a cost-minimising approach to estimate the projected uptake of different fuel types in line with the 2023 IMO strategy. Figure 3.1 presents the results. Assuming the adoption of a carbon levy, discussed in detail below, which is expected to close the price gap to current fossil fuel types, proposed by the EU Commission, demand for ammonia is expected to increase from 2027. From 2035 onwards, ammonia may be the dominant zero-carbon fuel type. It may drive the decarbonisation of the sector thanks to its cost efficiency, higher volumetric energy density, simplified production, and existing well-developed infrastructure compared to other alternatives (Hoang et al., 2022[6]). Ammonia is emission-free in combustion and can also be produced emission-free with green hydrogen (Box 3.2).

Figure 3.1. Estimated fuel mix for decarbonising international shipping

Estimation based on the revised 2023 IMO strategy and proposed measures by the European Commission

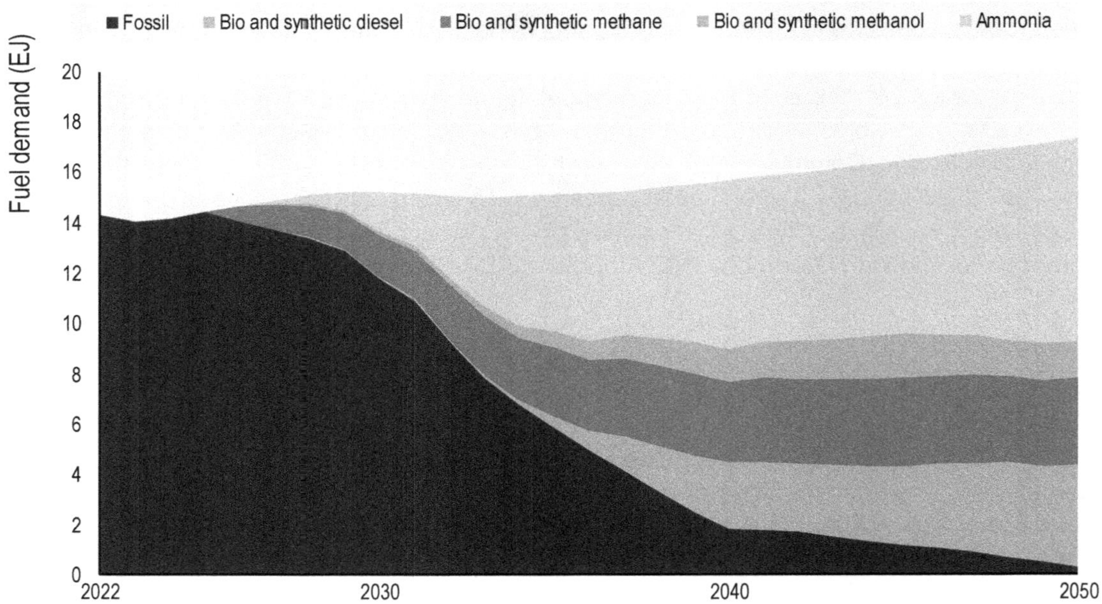

Note: Ammonia includes blue ammonia produced from natural gas with carbon capture and storage as well as green ammonia from green hydrogen.
Source: (Equitable Maritime Consulting, 2023[7])

StatLink https://stat.link/q7ckva

Box 3.2. Ammonia may be the most promising zero-emission shipping fuels

Ammonia is a cost-effective zero-carbon bunker fuel that is easier to produce and store in large quantities than other zero-carbon fuels. "Green" ammonia is produced from hydrogen made by hydrolysis of water with renewable electricity. Research and development efforts are required to address potential air pollution of nitrogen oxides and nitrous oxide from ammonia combustion. Since ammonia is toxic to humans and aquatic life it needs to be stored and managed safely.

Green methane and methanol can also be produced and used emission-free. They require carbon, which can be sourced emission-free in two ways (World Bank, 2021[8]):

- from sustainable biofuels, in biofuel synthesis (biomethane, biomethanol),
- as a synthetic fuel, through recovering CO2 from carbon capture and combining it with hydrogen, in hydrogenation for alcohol synthesis (synthetic methane, synthetic methanol).

Their advantage is that they are similar to conventional fossil fuels and therefore require relatively minor changes in conventional fossil bunkers and engines. However, the production of synthetic fuels is energy-intensive. The climate-neutral sourcing of captured CO2 may also be a concern. In particular, sourcing it from CO2 emissions in industrial processes would not be fully climate-neutral. Sourcing it from direct air capture would be climate-neutral but would increase energy needs. Biofuels should also be sourced sustainably to ensure their use is emission-free. Another limitation is the high demand for biofuels in multiple sectors to reach climate neutrality, as well as competing land use for agriculture or biodiversity protection.

Next to ammonia bio and synthetic methane and methanol, as well as advanced biofuels are also expected to see some uptake to reach the 20%-30% GHG reduction target by 2030. Safety is another important aspect of zero-carbon fuels (Box 3.3).

Box 3.3. Safety of zero-carbon fuels: The role of the Clean Marine Fuels Working Group

The Clean Marine Fuels Working Group (CMF) plays a crucial role in facilitating and regulating the supply of new marine fuels by providing expertise and guidance. CMF has taken significant strides in ensuring bunkering safety and broader system safety. Initially, a bunkering safety toolkit for liquefied natural gas (LNG) was introduced, and subsequently, it has been adapted to encompass all low- and zero-carbon fuels. On the operational safety front, comprehensive safety bunkering checklists for liquid gases were released in 2022, followed by one for methanol in 2023. Ongoing investigations are developing safety protocols for ammonia. Regarding system safety, CMF is currently in the process of transforming the existing "audit tool for bunker facility operators" and "bunker-ready terminal tool" into universally applicable tools suitable for all existing and emerging marine fuels.

The port can provide incentives for shippers' voluntary emission reduction

Ports can voluntarily offer incentives to promote and reward shipping companies for surpassing legal emission standards. To facilitate this effort, the International Association of Ports and Harbors (IAPH) established the Environmental Ship Index (ESI), which quantifies environmental performance in terms of air pollutants, CO2 emissions and noise. This index has been universally adopted by ports worldwide as a tool to incentivise ships to lower emissions below the IMO emission standard. Currently, the ESI database includes around 7,000 commercial ships and more than 60 organisations, primarily consisting of port authorities. The Hamburg Chamber of Commerce could encourage shipping carriers to make use of the Environmental Ship Index (ESI) and monitor emission reductions in line with its 2040 climate neutrality target.

Measures by ports to schedule the arrival and mooring of ships (port call optimisation) lead to GHG emission reduction through lower energy use. Effective cooperation is vital. To that end, the International Taskforce on Port Call Optimization (ITPCO) aims to achieve optimised port calls through collaboration among relevant agents (international shipping, ports, terminals, and cargo owners) thanks to timely high-quality data sharing and standardisation.

Establishing green shipping corridors involves identifying specific trade routes between major hubs that support zero-emission solutions. This initiative depends on voluntary cooperation among ports, shipping entities, and other stakeholders. Successful implementation will require the participation of ports from developing countries. The Clean Energy Marine Hubs (CEM-Hubs) platform, a cross-sectoral public-private initiative involving energy, ports, finance, and shipping sectors, aims to accelerate the production and maritime transport of low-carbon-emission fuels, including those directly used for shipping.

Modelling the cost impact of potential decarbonisation measures

This subsection discusses the scenario design and the modelling method used to assess the economic impacts of measures proposed by the European Commission to reach the 2023 IMO emission reduction and climate neutrality targets. The assessed impacts include trade costs and on the value of trade flows in Hamburg. They include a global fuel standard (GFS) and a carbon levy on ships with a capacity above 5 000 gigatons (GT).

In the proposed GFS emissions are assessed "well-to-wake" (WTW) across the whole value chain of fuels. WTW refers to emissions in the entire process of fuel production, delivery and use on ships, so comprises Scope 1, 2 and 3 emissions in fuel use. A 16% reduction in the carbon intensity of ship fuel is applied in 2030. According to the EC proposal, the GHG reduction from the GFS is raised to 80% and 95% by 2040 and 2050, respectively. In addition, a carbon tax of 150 USD per ton is applied by 2030, to be increased to 200 USD by 2050.

Scenario design

The economic impacts are assessed comparing a policy scenario with the proposed measures against a business-as-usual (BAU) scenario, which includes climate action on international shipping only up to the measures introduced by IMO in 2021. Both policy and baseline scenarios assume that global climate action more generally will limit global warming to 2 degrees (Box 3.4).

Box 3.4. The Business-As-Usual (BAU) scenario

The Business-As-Usual (BAU) scenario applies the Shared Socio-Economic Pathway (SSP) 2 and Representative Concentration Pathway (RCP) 2.6 to reflect global socio-economic development and climate actions expected to be taken by countries worldwide. SSPs are pathways that describe how global society, demographics and economic developments might change over the next century and are used as a backdrop to model climate action. In SSP2 trends broadly continue historical patterns. RCPs refer to atmospheric GHG concentrations and other forces acting on climate, resulting from climate action. RCP 2.6 is consistent with 2-degree global warming. The economic projections rely upon DG-ECFIN for the European Union, while the IMF and OECD long-term projections inform the outlook for non-EU economies.

Furthermore, the BAU scenario takes into account the International Maritime Organization's (IMO) short-term measures in 2021 —including the Energy Efficiency Design Index (EEDI), Energy Efficiency Existing Ship Index (EEXI), and Carbon Intensity Indicator (CII)— to curtail shipping carbon intensity (IMO, 2021).

EEDI gauges the ship designs energy efficiency, with lower EEDI values as an indication of increased ship efficiency. This index, introduced in 2013, applies to new ships. EEXI applies to ships' operating emissions. These two indices oblige ship owners to adhere to specific energy efficiency standards. CII quantifies CO2 emitted by ships during transport activity on an annual basis. It is specified in grams of CO2 emitted per cargo-carrying capacity and nautical mile. The IMO has ntroduced a rating system ranging from A to E with an increasing stringency towards 2030. Ships with a D rating for three consecutive years are required to apply a corrective action plan.

Methodology

There will be a direct and an indirect impact of the GFS and of the carbon tax on transport costs, given the strong interdependencies between transport and trade systems (Halim, Smith and Englert, 2019[9]).

First, a direct impact will be on ship running costs on account of fuel and capital costs, as well as on account of the carbon levy on remaining emissions. This will increase transport costs. The increase in these costs impacts global trade. In turn, transport costs are sensitive to the change in trade volume, as unit transport costs could respond to the loss or gain in economies of scale in shipping. Scale economies can reinforce trade-cost-induced shifts in trade flows.

To assess the economic impacts on trade through Hamburg, this study focuses on the following indicators:

1. Change in transport costs by commodity for imports and exports (measured in percentage change), based on a comparison between BAU and policy scenario;
2. Change in trade values by commodity for import and export (measured in USD and percentage change) based on a comparison between BAU and policy scenario;
3. Change in trade values for the top 20 import/export countries (measured in percentage change).

To assess these impacts, the analysis is carried out using the output of NavigaTE, described above and in the Annex. The approach also deploys Equitable Maritime Company's (EMC) International Freight Model (IFM) to model cost-minimizing international freight transport choices, including transport mode choices, taking into account the quality of infrastructure and shipping services at a detailed network level (Halim et al., 2018[3]) This analysis is complemented by the output of the GEM-E3 model, a global computable general equilibrium economic model designed to project worldwide commodity trade under different scenarios (E3modelling). The modelling approach follows three steps, illustrated in Figure 3.2. A description of the 3 models and the 3-step modelling approach is in the Annex. This analysis is complemented by the output of the GEM-E3 model, a global computable general equilibrium economic model designed to predict worldwide commodity trade volumes under different scenarios (E3modelling). A detailed explanation of these three models can be found in the Annex and the cited references. The modelling approach follows three steps, illustrated in Figure 3.2. A description of the 3 models and the 3-step modelling approach is in the Annex.

Figure 3.2. Methodology Diagram

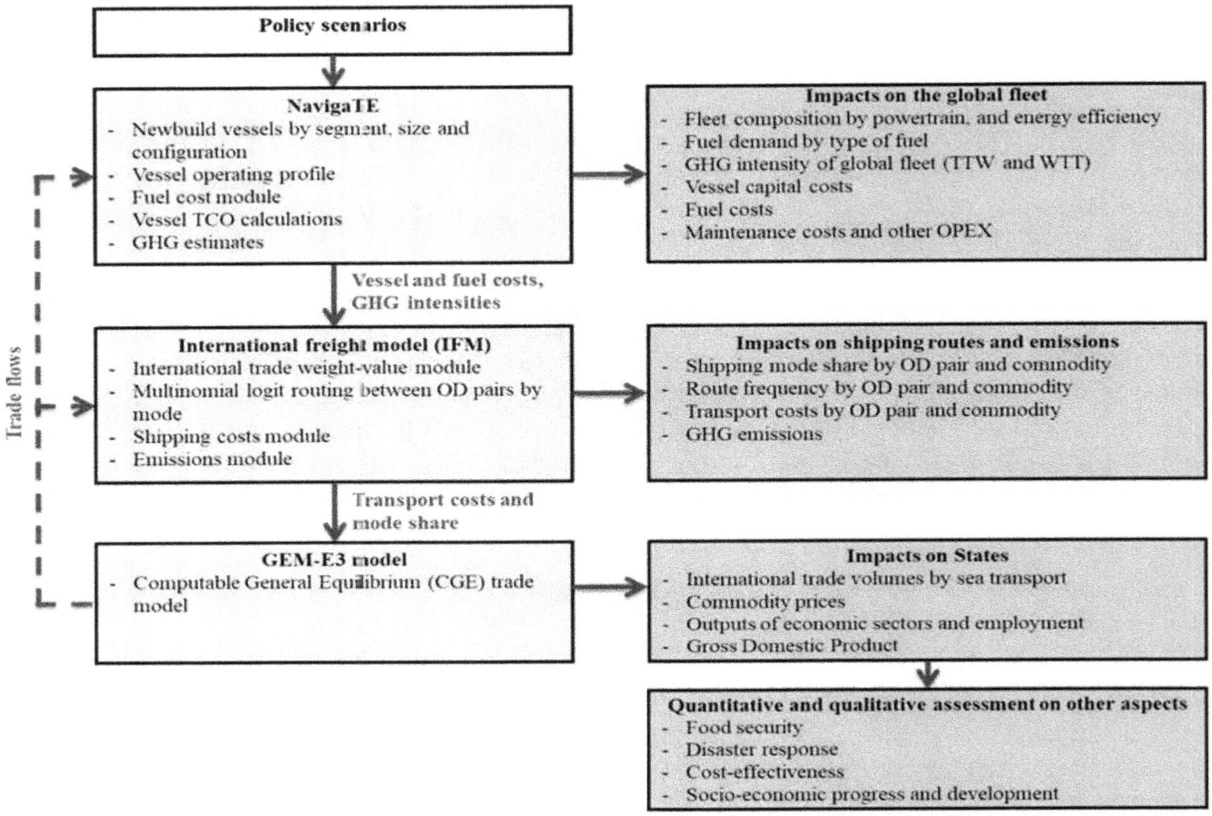

The impact of potential mid- and long-term decarbonisation measures: Modelling results

Impact on import and export costs by commodity

The GHG mitigation measures proposed by the EC may lead to higher ship running costs through higher fuel prices and higher capital costs. Users and producers of zero-emission fuels invest in new technologies and vessels, including advanced propulsion systems. Based on the assessment using NavigaTE, ship running costs are estimated to increase up to 20.37%, 17.78%, and 4.13% in 2030, 2040, and 2050 respectively, relative to BAU.

The estimations with the NavigaTe model suggest that the resulting increase in fuel costs due to the adoption of more expensive cleaner fuels is significantly higher than in vessel capital costs. However, this increase varies across ship types, routes, and commodities (Halim, Smith and Englert, 2019[9]). Even so, the measures will trigger a substantial increase in shipping company CAPEX around 2030 due to ship retrofitting and the purchases of ships with zero-carbon fuel capabilities.

While fuel costs may rise significantly initially, fuel costs are then projected to decrease gradually with the assumed improvement in the productivity of zero-carbon fuel production. Concurrently, lower emissions reduce costs on account of the carbon tax. Improvements in energy efficiency following regulation and technological progress, such as in hull shapes or propulsion systems, may also reduce costs. However, whether this benign long-term cost assumption materialises depends on many uncertain factors, including on the availability of renewable energy and green hydrogen world-wide, energy and hydrogen demand trends, as well as the choice of a cost minimising fuel mix, taking into account the needed energy inputs.

Table 3.1 presents the average impact on import unit transport costs of all commodities over time. By 2050, the increase in import transport costs across commodities may be relatively small. The changes in the top three commodities by import value for Hamburg are displayed in Figure 3.3 with broadly similar impacts. Across a broader range of commodity types (Annex Figure 3.A.1), import costs are estimated to increase between 20% and 32% in 2030. The highest costs concern raw materials which are mostly imported from developing countries.

Table 3.1. Projected average increase in unit transport costs across commodities imported in Hamburg

Impact of proposed IMO policy measures, relative to business-as-usual scenario

	2030	2040	2050
Average increase (%)	25,3	21,6	4,9

Source: (Equitable Maritime Consulting, 2023[7])

Figure 3.3. Projected increase in unit transport costs for top 3 commodities imported in Hamburg

Percent change in import unit transport costs (USD/ton-km), relative to business-as-usual scenario

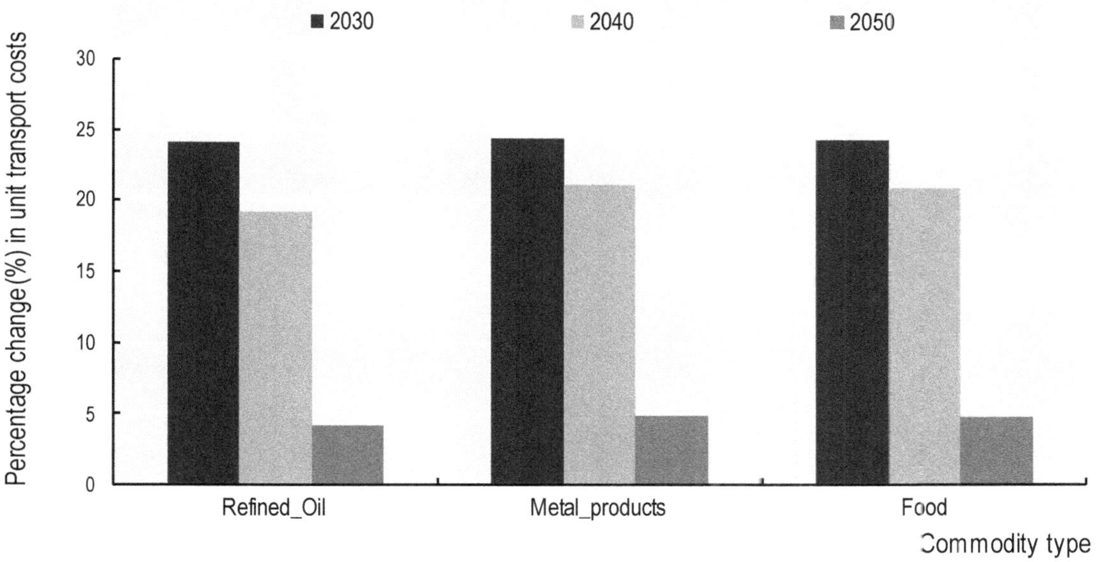

Note: Estimation based on the revised 2023 IMO strategy and proposed measures by the European Commission
Source: (Equitable Maritime Consulting, 2023[7])

StatLink https://stat.link/96fxcw

A similar pattern over time emerges for export unit transport costs (Table 3.2). Among the top 3 export goods for Hamburg increases in unit transport costs are bigger in the export of metal products and food than in transport equipment (Figure 3.4). They may also be somewhat bigger than in the case of unit import costs.

Table 3.2. Projected average increase in unit transport costs across commodities exported in Hamburg

Impact of proposed IMO policy measures, relative to business-as-usual scenario

	2030	2040	2050
Average increase (%)	28,5	24,5	5,5

Source: (Equitable Maritime Consulting, 2023[71])

Figure 3.4. Projected increase in unit transport costs for top 3 commodities exported in Hamburg

Percentage change in export unit transport costs (USD/ton-km), relative to business-as-usual scenario

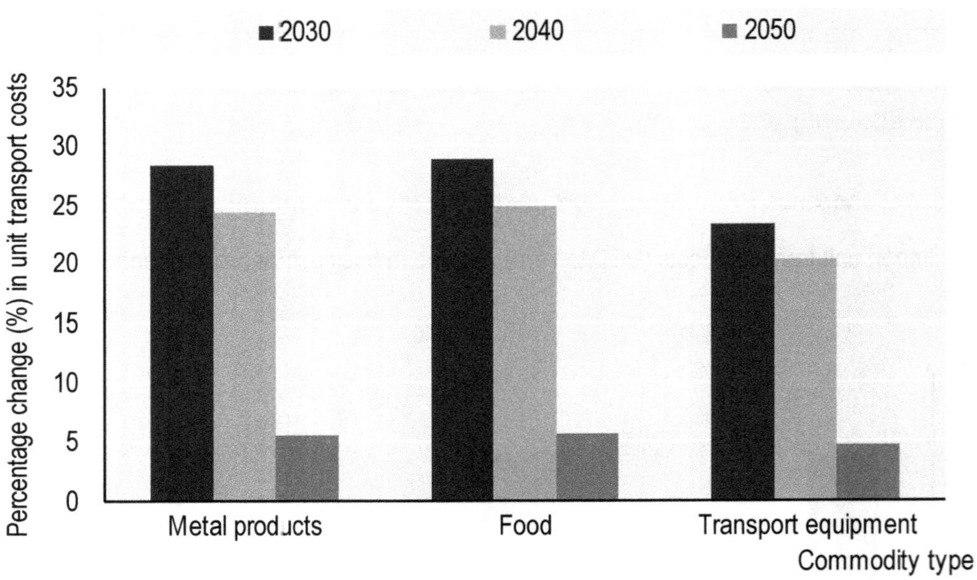

Source: (Equitable Maritime Consulting, 2023[71])

StatLink https://stat.link/jd14s6

Impact on import and exports by commodity

Higher transport costs could lower trade flows, for example as a result of firms relocating production. The projected impact on Hamburg trade flows is small (Tables 3 and 4). With its diversity of trading partners, Hamburg may have more potential to accommodate and adjust to changes in origins and destinations than other ports. Indeed, the port serves as a major hub in the maritime network, supporting this flexibility.

Table 3.3. Projected total Hamburg import values

Billion USD

Import value	Baseline	Policy scenario	Difference (%)
2030	63.4	62.8	-0.9
2040	67.4	66.9	-0.7
2050	71.7	71.6	-0.2

Source: (Equitable Maritime Consulting, 2023[7])

Table 3.4. Projected total Hamburg export values

Billion USD

Import value	Baseline	Policy scenario	Difference (%)
2030	54,9	54	-1,7
2040	59,6	58,8	-1,4
2050	66	65,7	-0.4

Source: (Equitable Maritime Consulting, 2023[7])

Food commodities may be subject to the most substantial decrease in both import and export values in absolute terms (Figure 3.5, Figure 3.6). Figures in the Annex present the percentage change in import and export values by commodity.

Figure 3.5. Projected change in import value of each commodity in Hamburg

Absolute change in import value (Billion USD), relative to business-as-usual scenario

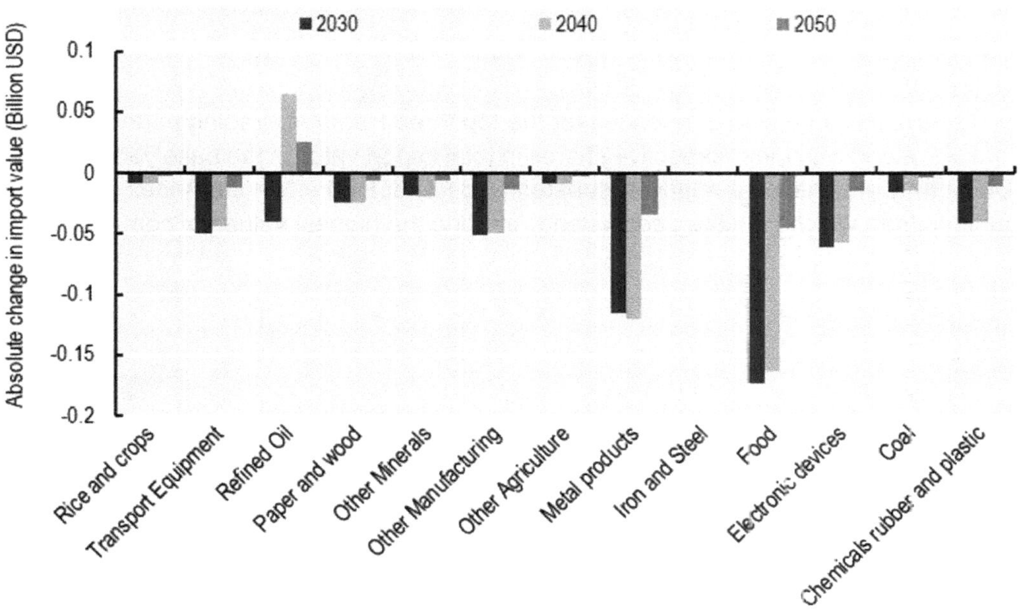

Source: (Equitable Maritime Consulting, 2023[7])

StatLink https://stat.link/i132us

Figure 3.6. Projected change in export value of each commodity in Hamburg

Absolute change in export value (Bi lion USD), relative to business-as-usual scenario

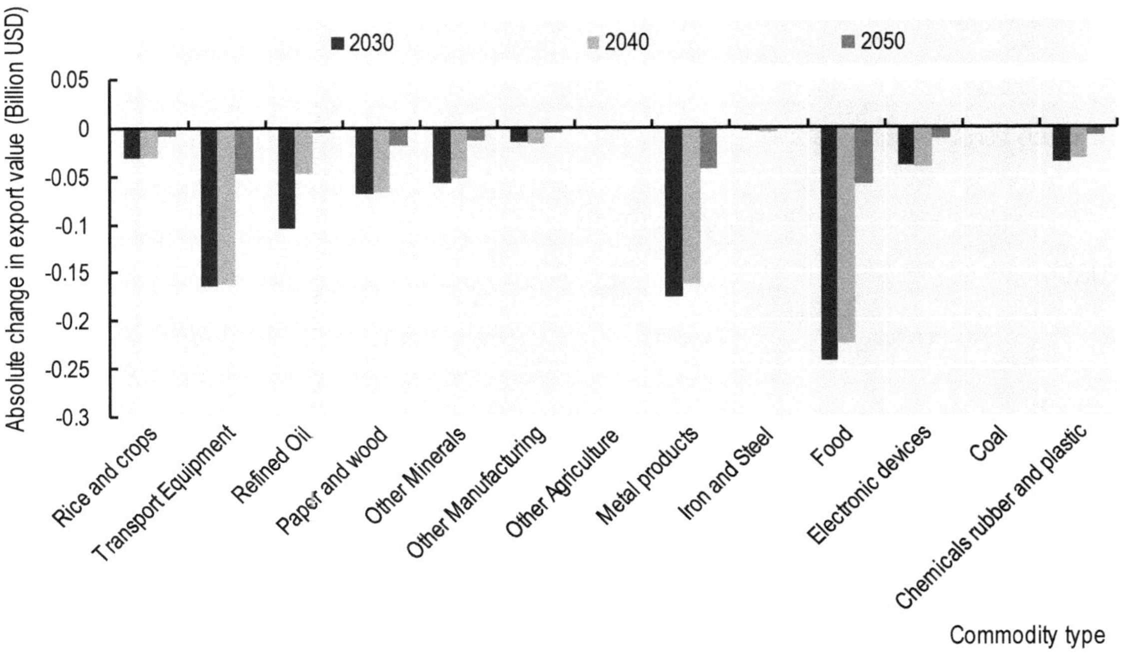

Source: (Equitable Maritime Consulting, 2023[7])

StatLink https://stat.link/zu8b6v

Impact on import and export value by country

Figure 3.7 displays the projected differences for the top three Hamburg trading partner countries in terms of import values, along with their respective shares in total import value in the base year (2022). The results suggest China may experience a relatively substantial decline. Figure 3 in the Annex shows the impact for the 22 countries from which Hamburg consistently imports the highest values of commodities.

Figure 3.7. Projected change in imports to Hamburg from the top 3 trading countries

Percent change in import from the top 3 countries, relative to business-as-usual scenario

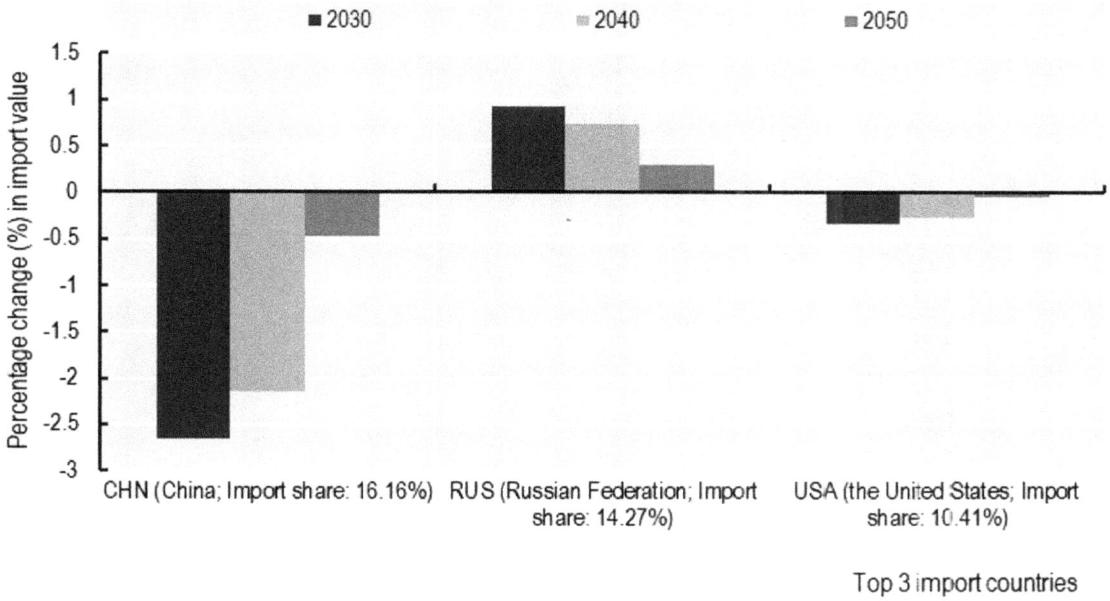

Source: (Equitable Maritime Consulting, 2023[7])

StatLink https://stat.link/tbze3x

Figure 3.8 displays the changes in export value for the top three countries in the base year. The results suggest all of them experience a decline, notably China. Figure 4 in the Annex presents the analysis for the top 20 countries.

Figure 3.8. Projected change in exports from Hamburg to the top 3 trading countries

Percentage change in export of Hamburg to top 3 countries, relative to business-as-usual scenario

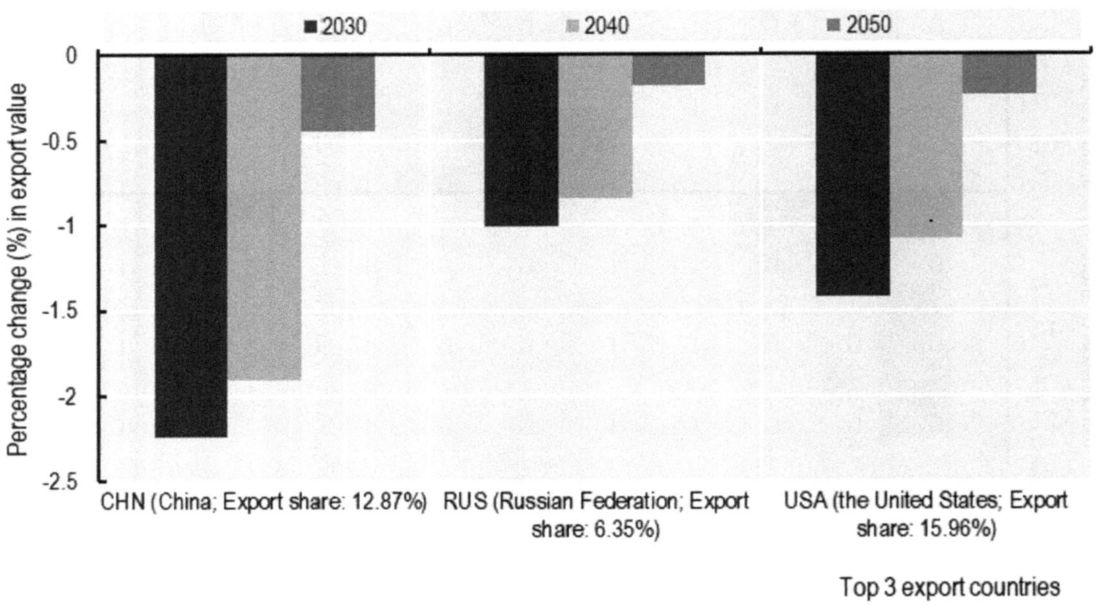

Source: (Equitable Maritime Consulting, 2023[7])

StatLink https://stat.link/w9emc4

Key actions

Immediately

- The Hamburg Chamber of Commerce could, together with the port authority, key shippers in Hamburg, as well as researchers, and in coordination with other major ports, assess advances in the deployment of zero-emission fuel and ships, with a view to promoting fuels likely to do best in terms of system-wide low cost and resilience.
- The Hamburg Chamber of Commerce could encourage shipping carriers to make use of the Environmental Ship Index (ESI). It could provide incentives for shippers' voluntary emission reduction, by rewarding companies that surpass legal emission standards.
- All new ships should run on, or be able to run on, zero emission fuels, ideally anticipating a long-term system-wide low-cost fuel mix.

By 2030

- Shipping companies could augment zero-carbon fuel use, such as ammonia. Ensuring safe production, transport and delivery of zero-carbon fuel will be key to that end.
- The Hamburg port could cooperate with other major ports and shipping entities to establish green shipping corridors to identify specific trade routes between major hubs that support zero-emission solutions.
- The Hamburg Chamber of Commerce could continue assessing the impacts of transport costs on trade flows through Hamburg.

By 2040

- Broadly complete the replacement of fossil fuels with zero-emission fuels in international maritime shipping.

Climate neutrality in transport logistics and transport from and to the port

Hamburg is the largest rail port and the busiest road freight transport node in Europe, and the third largest inland water transport (IWT) port in Germany. One in seven of all rail freight journeys in Germany start or finish in Hamburg, carrying 15% of Germany's rail freight. Some 200 freight trains use the port daily. Hamburg is the largest rail container transhipment centre in Europe. Three out of nine major European road freight transport routes – the TEN-T corridors, which account for 80% of EU road freight transport – intersect in Hamburg, which is by far the busiest node, with the Hamburg-Berlin route alone accounting for over 9 Mt annually (Figure 3.9). Hamburg is one of the largest road freight loading regions in Europe (OECD, 2021), with 40 000 trucks driving into the port daily. With good connections to the pan-European canal and river network, the port also attracts 11 000 calls annually from inland vessels travelling essentially to Berlin, Hannover and the Czech Republic.

Just over half of containers were transported by rail (50.5%), road accounting for 47.3%, and inland water transport for 2.2%. Some 53.9% of freight tonnage was by rail, 37.6% by road and 8.5% by inland water transport (Port of Hamburg Marketing, 2023[10]). In other large ports, the share of rail ranges from 7% (Le Havre and Antwerp) to 11% (Rotterdam) (Figure 3.10). Only two smaller ports reach a higher share. Inland water transport covers 9% of hinterland freight transport (mainly bulk cargo) and 2% of container transport out of Hamburg, whereas in the ports of Rotterdam or Antwerp, inland water transport accounts for half of freight tonnage and 40% of container transport.

The Port of Hamburg stands out among European ports because of the large share of rail in hinterland transport, which has almost doubled over the last 20 years, largely at the expense of road. Four main factors have helped this increase:

- Hamburg has been able to leverage its position as a gateway for long-distance trade thanks to good rail connections (2 000 offered daily out of the port) and a hinterland network that could handle a large number of trains with segments with third tracks and sidings for longer trains.
- An extensive railway system within the port gives rail a head start from the moment of unloading. Ports that have high shares of rail for hinterland transport, such as Gothenburg, Trieste and Koper, share this characteristic. Over half the HPA rail track is electrified, a higher share than other major ports in Europe (for instance Antwerp and Rotterdam).
- There have been major investments to remove infrastructure bottlenecks within and through the port, such as the Kattwyk railway bridge, one of the largest in the world.
- The governance structure of the port and a high degree of vertical integration have supported rail choices, for instance opening up HPA tracks to all 57 railway operators.

Figure 3.9. Road freight traffic over the main Ten-T corridors (2020)

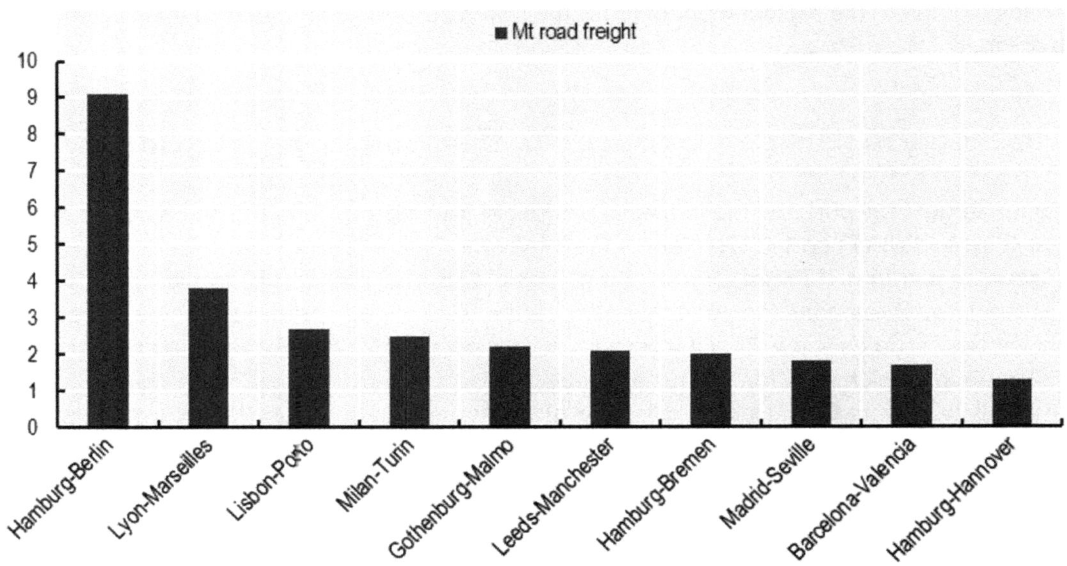

Source: (T&E, 2021[11])

StatLink https://stat.link/0emqhc

Figure 3.10. Hinterland container transport from European ports by modal share

Thousand twenty-foot-equivalent container units (TEU), 2022 or the latest available date

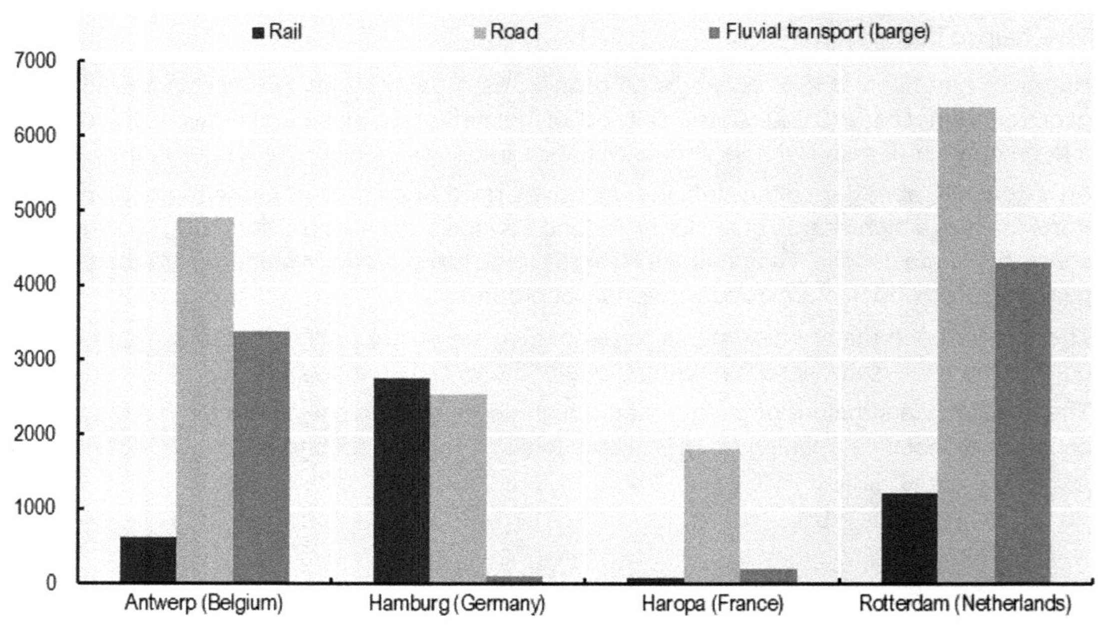

Source: Port authority reports

StatLink https://stat.link/cri0a7

The high share of rail in hinterland transport places Hamburg in a favourable position to work towards climate neutrality since much of its freight transport is already zero-emission. This share is concentrated in long-haul transport (up to 80% of long-haul freight is transported by rail) so that much of road transport takes place on shorter trips, a segment more amenable to electrification.

Hamburg is also a major logistics hub, combining short, medium and long-range transport with warehousing and storage services. The logistics industry occupies a significant part of Hamburg's commercial real estate, covering one-third of its total warehousing and storage area. Urban freight logistics are a growing segment of the logistics business, bolstered by Hamburg's high population and value-added density. In 2019, the average number of courier, express and parcel items per inhabitant in Hamburg was a third above the national average, with related traffic already accounting for about 10% of inner-city traffic.

Acting on Hamburg's competitive advantage: governance, vertical integration and interlinkages in hinterland transport and logistics

The Port of Hamburg is an intermodal hub serving shippers and freight forwarders who are clients of a multitude of hinterland transport and logistics companies. Transport choices are often part of multimodal supply chain strategies, especially as containerisation has reduced the time and cost of transferring cargo.

The port, city transport and logistics businesses are closely interlinked through cross ownerships, partnerships and long-term commercial relationships that support the use of rail for hinterland transport:

- The City of Hamburg is the majority shareholder of the HPA and of the largest port terminal operator HHLA.
- Both terminal operators (HHLA and Eurogate) hold substantial equity shares in railway operators (Metrans and EUROKOMBI respectively) and rail terminals in the hinterland. EUROKOMBI is Germany's largest intermodal railway terminal. Metrans links its port terminals to inland rail terminals in Germany, Poland, Hungary, Slovakia and the Czech Republic.
- The Port of Hamburg is one of Deutsche Bahn's largest customers. In addition to Transfracht International (TFG), companies such as Railion, Intermodal, HHCE, Interconainer-Interfrigo or Kombiverkehr also transport goods straight from the port to the hinterland by rail.
- Some rail freight operators, such as DB Cargo, are also active as third-party logistics companies and freight forwarders.

The actors in road hinterland transport are more dispersed, with 1 700 road haulage companies operating 45 000 trucks registered with the Hamburg Chamber of Commerce. Over 120 courier services are available in Hamburg for the transport of small and urgent consignments across Germany and beyond. Some are also active in the urban logistics business in Hamburg.

In addition to these specialised trucking and freight rail operators, combined transport operators offer intermodal services. For instance, the Hamburg-based Zippel Group has a fleet of 200 trucks and operated 2 000 block trains in 2019. Third-party logistics service providers offer storage, packaging and customs brokerage, and a range of door-to-door hinterland transport services. Container lines such as Maersk, CMA CGM, Cosco and NYK, have their own freight forwarding subsidiaries. Some freight forwarders (such as Kühne+Nagel) hold shares in shipping companies (in this example, Hapag Lloyd). DB Schenker, a division of Deutsche Bahn, is a global logistics provider.

Altogether, about 12 600 companies employ 400 000 people in the local logistics industry. Shippers range from small businesses to huge multinationals. Retail contributes to a large part of containerised shipping and procures considerable amounts of freight transport services directly or via freight forwarders. Larger retailers can own warehouses, vehicles for last-mile transport, or even their own freight forwarding companies, such as Hermes, which belongs to the Hamburg-based retailer the Otto Group, one of the world's largest e-commerce businesses.

Breaking the link between growing freight activity and CO_2 emissions

At 3 602 Mt and 417.2 billion tonne-kilometres, German freight transport is by far the largest in Europe, accounting for almost one-quarter of total EU tonnage, and twice as much as the next largest (France and Spain). The share of rail has grown from 16% to 19% over the last twenty years because the fall in traditional bulk cargo has been more than compensated by an increase in container transport, thanks to substantial investment at key connection points such as ports (BALM, 2023[12]). By contrast, in Europe, the share of freight transported by rail has halved to about 15% in the last 40 years, prompting a downward cycle of increasing fixed costs, loss of competitiveness and loss of volume.

Road freight transport in Germany reached 303.50 billion tonne-kilometres in 2022, 89% of which was national transport. Over half of road tonnage is transported less than 50 km. The German fleet of trucks numbers about 2.7 million, with close to one million vehicles over 3.5 tonnes. Some 85 000 heavy-goods vehicles (HGVs) are newly registered every year.

At 5.1% of total freight transport, Germany has one of the largest modal shares of inland water transport in Europe. Some 182.45 Mt of goods were transported on German inland waterways in 2022, of which over 75% was international transport. Around 80 % is transported on the Rhine for at least part of its journey.

Projections of the effect of current policies agree that inland freight transport will grow by at least 30% into the 2040s, and that road transport will grow more strongly, reinforcing its dominant position in the modal mix:

EU forecasts show an increase of 31% in inland freight transport by 2030 compared with 2015, and of 55% by 2050, with road freight transport growing by 35% by 2030 (ITF, 2021[13])

In Germany, a recent forecasting exercise for the Federal Ministry of Digital Affairs and Transport (BMDV) sees freight transport grow by 30% and road freight transport by 34% by 2051 compared with 2019 (Intraplan Trimode, 2023[14]).

In 2019, transport accounted for 23% of Germany's greenhouse gas (GHG) emissions. Road transport represents 80% of total transport emissions (196 MT of CO_2 equivalent), of which around 28% are from heavy-goods vehicles (UBA, 2023[15]) If projections for freight activity turn into reality, without decarbonisation, freight transport would continue to contribute substantially to CO_2 emissions. EU and national emission reduction objectives therefore appear relatively ambitious:

- The EU aims for a 55% reduction in GHG emissions by 2030 compared with 1990 and climate neutrality by 2050, with a 90% reduction in CO2 emissions from transport, and a 50% increase in the share of rail in freight transport by 2030 and a doubling by 2050.
- Germany aims to reduce its GHG emissions by at least 65% by 2030 compared with 1990 levels, reach net-zero emissions by 2045, electrify 30% of HGV vehicle-kilometres and increase the share of rail in freight transport to 25% by 2030.

The HPA is aligned with the Hamburg Chamber of Commerce's objective of climate neutrality by 2040. The ports of Antwerp and Rotterdam have set this objective for 2050. The HPA has also set a target of 65% for hinterland freight transport by rail by 2040, as well as a target for train container capacity utilisation of 80 TEU in 2025 (BWI, 2023[16]). HHLA has signed up for climate neutrality by 2040 concerning its operations (scope 1 and 2) and aims to reduce its emissions by 50% by 2030, compared with 2018 (HHLA, 2023[17]). Some larger transport operators have formulated emission reduction targets. For example, Deutsche Post DHL Group has set itself a 2050 climate neutrality objective, with a 2025 target to increase carbon efficiency by 50% compared with 2007. Several large shippers have carbon emission reduction targets that cover both their own operations and those of their transport and logistics service providers. For instance, the Otto Group aims to be carbon-neutral by 2030 for its transport and logistics, and IKEA aims to reduce its carbon footprint from freight transport by 70% and from logistics by 80% by 2030.

Opportunities and barriers on the way to climate neutrality

Reconciling the projected growth of freight transport with ambitious climate neutrality targets will require a major effort to decarbonise and improve energy efficiency in freight transport and logistics. Options relevant to Hamburg and its hinterland include the following:

- Modal shift from road to rail
- Modal shift from road to decarbonised inland water transport
- Options for the decarbonisation of road transport
- Energy efficiency improvements in hinterland transport and logistics

Their potential contribution to making Hamburg's economy climate-neutral by 2040 is evaluated in this section, in terms of their costs, their effectiveness and their timeframes.

The potential for further modal shift to rail

Shifting freight from road to rail depends on two factors: making rail more attractive on the one hand and securing additional rail capacity to handle the increased demand on the other.

Distance is one of the main differentiators in mode choice: longer distances bring economies of scale and flow bundling opportunities, and reduce intermodal handling cost relative to total cost. Weight and volume are also key factors, with small volume and low-weight goods such as textiles and food products frequently shipped by road, but over 80% of coal and crude oil transported by rail or inland water transport. Rail is the most competitive transport mode for high volumes and weight, long shelf life and low sensitivity to transport conditions, over longer distances, the sweet spot being generally beyond 300 km to 500 km. Ports such as Hamburg have the advantage that they can generate the critical mass of cargo needed to operate high-frequency large-capacity shuttle trains to the hinterland.

A recent ITF study found that because rail is already very cost-effective for long-distance freight transport, when it already accounts for a large share of freight transport as is the case for hinterland transport out of Hamburg, generating modal shift is difficult (ITF, 2022[18]). The main factor blocking change is inelastic demand for road transport: the reach, speed, and flexibility of road transport are generally superior to rail, so price changes do not always alter the mode-choice decisions of shippers. Modal shift efforts therefore need to focus on increasing attractiveness beyond cost considerations, notably by encouraging multimodal transport and reducing intermodal dwell time. The modal shift requires better information so users can make the appropriate choices. Digital technology is critically important in this respect, starting in ports. The Port of Hamburg's EVITA/TransPORT Rail digital platform includes rail information and plays an important part in supporting rail hinterland transport.

If demand for rail shifts from road transport, rail capacity will need to grow. Resolving rail bottlenecks around Hamburg involves substantial investment with long lead times. For instance, the Hamburg to Hannover corridor is at full capacity and in need of renovation, and the Hamburg to Bremen connection needs upgrading and a third track. Dry ports or extended gates have reduced bottlenecks and increased rail use in ports such as Gothenburg and Antwerp, but are not an option for Hamburg due to spatial constraints and lead times beyond 2040 (Merk and Notteboom, 2015[19]). A further issue is that in Europe, passenger rail transport is prioritised over freight, restricting access to daytime slots, especially close to critical nodes. Increasing freight capacity would require dedicated freight lines, many of which have been dismantled to cut costs. Recreating such redundancies would be a lengthy and, in some cases, an impossible task, since land has long been sold off.

It is generally considered that the most realistic solution to capacity constraints is to focus on making the best of currently available routes through large-scale digitalisation: advanced train control and signalling systems using wireless communication to supervise trains could increase capacity by more than 20% on many network lines without additional tracks (ITF, 2022[20]). The 2000 European Rail Traffic Management

System legislation has led to technological improvements, but progress has been slow, partly because digitalisation requires transformation rather than a step-by-step change. Except for a few countries, the pace of digitalisation has been slow, including in Germany. The Digitaler Bedienplatz should update a portion of the system, though completion is only scheduled for 2033 to 2035.

Modal shift to decarbonised inland water transport

Cargo transport by barge emits up to ten times less CO_2 per tonne than by road (and up to 5 times less than rail). However, reaching climate neutrality requires switching from barge diesel engines to zero-emission propulsion, which is a big challenge. Inland water transport shares structural characteristics with rail in terms of cost, time and flexibility. It also suffers from infrastructure restrictions, notably the bottleneck in the Scharnebeck ship lift south of Hamburg, as the new lock will not be completed until the early 2030s. The Elbe and the Rhine regularly struggle with low water levels, which are affected by climate change. In addition, within the port itself, access to barges will need to be expanded to increase capacity.

Zero-emission barges are being tested in Hamburg. A hydrogen fuel cell-powered pusher boat and a battery electric powered workboat are being tested in the port (Port of Hamburg Marketing, 2023[10]). A feasibility study for the City of Hamburg of an electrified barge for last-mile transport did not show significant energy use and emission reductions compared with electric trucks for loads under its maximum tonnage (108 tonnes), thereby reducing its usefulness for urban short-range transport of small loads (Fraunhofer Institute, 2022[21]).

The inland water transport industry is small and fragmented. The technological leap needed to move to zero-emission inland water transport is likely to require progress made with electric propulsion in sea shipping and road haulage. The uptake of new technologies will be slowed by the large variety and long lifespan of vessel hulls and engines, which exceeds 30 years. Options for totally decarbonising inland water transport therefore appear currently limited.

Decarbonising road freight transport and logistics

According to the BMDV, reducing transport CO_2 emissions by up to 48% by 2030 will entail one-third of the mileage of road freight transport being zero-emission, and one-third of semi-trailer trucks (about 145 000 vehicles) being zero-emission vehicles. These are transformative challenges. The main options for decarbonising road freight transport in the 2020s and 2030 are generally considered to be:

- Battery electric vehicles (BEVs)
- Electric road systems (ERSs)
- Hydrogen-powered fuel cell electric vehicles (FCEVs)

Though these options have been extensively analysed, major uncertainties remain as to their potential, if only because scaling up involves overcoming many technological, market and policy barriers.

Battery electric vehicles

With over one million electric cars on the road, the German electric passenger vehicle market has matured. BEVs are now also progressing in the light commercial vehicles (LCV) segment: according to the German Association of Automotive Industry, there were over 180 000 battery electric light commercial vehicles in Germany in 2021, up 6% from 2020 (VDA, 2022[22]). Much of this increase is attributable to last-mile delivery vehicles and short-haul trucks, which have predictable daily range and payload, return-to-base operations and charging.

Trips over 400 km make up around 5% of all trips in Europe but represent 40% of the EU truck activity (in tonne-km) and 20% of truck emissions, with similar figures for German road freight. BEVs with up to a 500 km range are entering the market. Charging during the driver's mandatory rest period (generally

45 minutes every 4 and a half hours) can extend the range to cover over 90% of road freight activity in Germany (T&E, 2021[11]).

The electrification of vehicles above 7.5 tonnes will require a high-power charging infrastructure. Battery electric trucks that use high-power fast charging need smaller batteries, shifting the economics of battery electric heavy-goods vehicles (ITF, 2022[23]). The numbers for the infrastructure needed are large: the European Automobile Manufacturers Association (ACEA) has estimated that heavy-goods vehicles would require up to 279 000 charging points across Europe by 2030, of which 84% would be in fleet hubs and the rest mostly public high-power points along highways and in overnight charging points (ACEA, 2022[24]).

As part of the EU "Fit for 55" package of regulation, the 2023 revision of the Alternative Fuels Infrastructure Directive mandates that 15% of the entire TEN-T be equipped with fast-charging stations at least every 120 km by 2025, increasing to 50% by 2027, and 100% by 2030, when the maximum distance between stations will be 60 km in the core TEN-T and 100 km in the comprehensive TEN-T (ICCT, 2023[25]). In Germany, the 2022 Federal Master Plan for Charging Infrastructure II proposes 10 measures to accelerate the expansion of HGV charging infrastructure, primarily information-sharing for mapping demand and grid requirements, but also tendering for a HGV fast-charging network along main transport axes (BMDV, 2022[26]). Business initiatives include a partnership of 20 research institutions and businesses, including MAN and ABB, which is working on the publicly funded "HoLa" megawatt high-performance charging system. Daimler Truck, the Traton Group (Volkswagen) and the Volvo Group have also come together for the Milence project, which is building two-megawatt charging systems along the A2 highway.

Electric road systems

An alternative to static charging is an electric road system (ERS) with an overhead catenary line which can also recharge a truck's battery, storing enough power to drive short distances. In Germany, Electric road systems (ERSs) are being tested with Federal funding on two short highway sections and a national road, with results expected at the end of 2024. Siemens, which is testing the concept in Germany and Sweden, estimates that 4 000 kilometres of ERS could accommodate about 60% of German truck traffic on the busiest routes.

Despite high energy efficiency, the high upfront cost of overhead cables (about EURO 10 billion for 4 000 km) appears to be a major barrier to adoption (Fraunhofer Institute et al., 2022[27]). The risk is that uptake might be limited by market inertia, or that ERS technology be rendered obsolete by BEV improvements. There are also major organisational obstacles to be overcome, such as reaching a Europe-wide agreement on technical standards.

Hydrogen-powered fuel cell electric vehicles

Hydrogen-powered FCEVs are less energy efficient than BEVs, but for heavy-goods vehicles driving long distances, weight is a game-changer: hydrogen is significantly more energy-dense. For an 800 km range truck, the weight difference can reach 2 tonnes. FCEVs can travel further with heavier payloads with shorter layover time since refuelling is generally faster than recharging. The range advantage means FCEVs would be suited for trips over 1 200 km, though such trips make up less than 9% of tonne-kilometres in Germany.

There is considerable interest in Hamburg around the hydrogen economy. A 100 MW electrolysis plant is being built on the site of the former Moorburg coal-fired power station. By 2026, it will provide green hydrogen for industrial processes, and transport and logistics (REH, 2023[28]). Ultimately it could be scaled up to 800 MW. A study for the logistics company Dachser identified Hamburg as one of 4 strategic locations where FCEVs could initially operate due to the availability of hydrogen and favourable location (Dachser, 2022[29]). The regions of Hamburg and Lower Saxony have launched an EUR 32 million project to replace diesel trucks with FCEVs. The Clean Cargo Connect project will build five hydrogen refuelling stations and two mobile refuelling facilities, as well as an electrolyser close to Oldenburg.

According to a recent ITF study, the key challenges are energy conversion losses, high vehicle cost (over EUR 400 000 compared with around EURO 250 000 for battery electric trucks and EUR 130 000 for diesel trucks), refuelling infrastructure and the price of hydrogen (ITF, 2023[30]). Fuel cells and hydrogen will likely not see substantial cost reductions from scaling up this decade. Other fuel cell markets such as maritime shipping might not start growing significantly before the 2030s.

Improving energy efficiency in hinterland transport and logistics

Energy efficiency improvements will be essential during the transition to climate neutrality and beyond, since demand for sustainable electricity will increase for a broad range of energy uses. For instance, charging the 40 000 heavy-goods vehicles that arrive in the Port of Hamburg daily would require as much as 16 GWh (for 40-tonne trucks with a battery capacity of 400 kWh). While they would not be charging simultaneously, this would require careful capacity planning. The upper-bound estimate of fuel efficiency improvements expected from the recent revision of European standards would reduce truck fuel consumption by 21% by 2030 compared with 2020. This level of efficiency would require an optimised aerodynamic tractor design, which would also reduce the electricity consumption of electric heavy-goods vehicles.

Reducing the energy intensity of freight transport also relies on route optimisation and reducing empty runs. In 2022 in Europe, 20% of all road freight kilometres were travelled by empty vehicles, with a higher share for national transport (24%) than for international transport (13%). About 20% of trucks had suboptimal loads (European Commission, 2022[31]). In Germany, the LKW Maut highway toll is estimated to have reduced empty runs by 2% (T&E, 2017). There remains scope for optimisation, especially for smaller transporters and for light commercial vehicles that will start paying the LKW Maut in 2024.

The Port of Hamburg is positioning itself as a major hub for e-commerce freight flows. E-commerce brings the challenge of managing urban logistics, with consignments in Hamburg expected to grow by 71% to 163 million by 2030. Hamburg's 2020 Sustainable Mobility Plan includes a strategy for decarbonising urban logistics, with a 40% emission reduction target for the last mile by 2030 compared with 2017. Hamburg is a testbed for innovative solutions to handle the sharp increase in deliveries with decarbonisation and efficiency improvements. Initiated by Logistics Initiative Hamburg (LIHH), efficient delivery networks include micro-depots instead of large storage centres, floating depots on unused canal sections and shared depots, vehicles and lockers (LIHH, 2023[32]). A trial with UPS found that 100 micro-hubs in the city centre could shift 40% of last-mile deliveries to cargo bikes. Other initiatives include Digital Hub Logistics Hamburg (one of 12 digital hubs chosen by the Federal Ministry of Economics and Energy to support digitalisation), which connects companies, startups, investors and researchers. The Hamburg Ministry of Economic Affairs and Innovation is assessing the feasibility of shifting courier traffic to urban waterways with support from the EU DECARBOMILE project. LIHH and nine partners are also testing inland water transport as part of the European sustainable urban freight AVATAR project.

Logistics is more than mobility: it includes warehousing and storage, where the main measures to improve energy efficiency and reduce emissions are equipment electrification, LED lighting and adapting the layout to reduce forklift and truck movement. For instance, all the forklifts operated in Kühne+Nagel warehouses are now electrified, and 75% of lighting is LED. With the growth of e-commerce, logistics real estate is expanding, providing opportunities to build sustainably, as for a new 53 300 m2 logistics centre in Hamburg Wilhelmsburg, with geothermal heating, 6 000 m^2 of solar panels to power BEVs, and state-of-the-art insulation, heating and air exchange systems to minimise energy use.

Moving faster towards climate neutrality: The effect of policy measures

"Peak internal combustion engine" for passenger cars is in sight and the electrification of road freight is underway: battery electric light commercial vehicles produced at scale can already be cost-competitive with diesel vehicles, given current battery prices (ITF, 2020[33]). The electrification of larger vehicles, because they are heavier and travel longer distances, requires expensive high-power chargers. According to ITF analysis, battery electric heavy-goods vehicles are not likely to reach total cost of ownership parity with diesel trucks until 2037. As for FCEVs, they would only be cost-competitive with hydrogen priced below EUR 2.5/kgH_2 (at the pump), down from the current EUR 11/kgH_2, which represents a major challenge (ITF, 2022[18]).

The cost-competitiveness of zero-emission technologies depends largely on economies of scale over the coming decade. Batteries are experiencing a self-reinforcing dynamic driving down costs in the passenger car market. This dynamic is spilling over into the urban and regional delivery truck segment. If it takes another 15 years to do so into long-haul trucking, it would be too late. Competitiveness parity should be reached before 2030 because of slow fleet turnover: in Germany, the average age of light commercial vehicles is 8 years, and 9.5 years for heavy-goods vehicles (Eurostat, 2023[34]). According to ITF's 2023 Outlook, assuming battery electric light commercial vehicles are already competitive with diesel vehicles and that battery electric heavy-goods vehicles become so in 2030, their share of the fleet by 2040 would only be 30% to 60% for light commercial vehicles and 15% to 30% for heavy-goods vehicles (ITF, 2023[35]).

Germany has introduced one of the highest purchase subsidies for zero-emission trucks in Europe, covering up to 80% of additional vehicle costs and/or charging infrastructure, with an EUR 500 000 cap, but take-up has been small. The 2023 revision of the European HGV CO_2 standards raises targets for manufacturers of zero-emission trucks to 45% of their sales from 2030, 65% from 2035 and 90% from 2040, in line with the EU Green Deal objective to decarbonise by 2050. They are designed to incentivise manufacturers to ramp up production to avoid supply bottlenecks in the zero-emission vehicle market.

The 2022 amendment to the EU Eurovignette introduces a tax of EURO 200 per tonne of CO_2 for HDVs in Member Countries with public distance-based tolling, with an optional higher external-cost charge for CO_2 emissions limited to 16 cents/km in all or part of their highway network. Germany's highway toll system, the LKW Maut, introduced in 2005, applies to all vehicles over 7.5 tonnes (extended to vehicles over 3.5 tonnes in 2024), except zero-emission vehicles. Its rate ranges from 9.8 cents/km to 35.4 cents/km, depending on vehicle weight, axle number and EURO emission category (Toll Collect, 2023[36]).

Taking action for the Hamburg economy in the lead-up to climate neutrality

Achieving climate neutrality by 2040 requires a proactive attitude on the part of all stakeholders in the Hamburg economy. This calls for a dedicated plan of action by the Chamber of Commerce and by businesses active in hinterland transport and logistics.

Actions for the Hamburg Chamber of Commerce

The first step towards climate neutrality is to shift from an emission reduction strategy based largely on transition fuels, with an energy mix centred on LNG and CNG, biofuels, and grey or blue hydrogen, to one focused on zero-emission solutions relying on renewable electricity. Renewable electricity generation has long been a major feature of the city-state: 28 offshore wind farms now dot the North Sea and Baltic Sea, several of which are managed from Hamburg. The city already prioritises deep geothermal and waste heat for thermal uses so that valuable renewable electricity can be directed to transport and other hard-to-abate uses. All stakeholders should integrate the need to manage this electricity efficiently into their climate neutrality strategies.

The Port of Hamburg has been remarkably successful in promoting rail for hinterland freight transport, contributing to a virtuous circle that helped preserve the overall share of rail in Germany's freight transport: according to a study by McKinsey, targeted rail infrastructure investment such as that by the Port of Hamburg have triggered larger shifts to rail than investments dispersed across European rail networks (McKinsey & Company, 2022[37]). The City of Hamburg has voiced support for the Federal goal of increasing the share of rail freight transport from 18% to 25% by 2030 and is keen to address the bottlenecks on railways out of Hamburg. This will require coordination with different tiers of government, including in non-transport policy areas such as spatial planning.

The city and the port benefit from a major competitive advantage in the transition to climate neutrality, since so much of hinterland freight transport is already zero-emission thanks to rail. The large share of rail, especially for long-haul trips, will provide a further advantage in decarbonising road freight transport: battery electric trucks are moving fast to cost parity with diesel in the short to medium-distance segment; for long haul trucks, this might not happen before the 2030s. In the case of Hamburg, with rail well established on long-haul routes, decarbonisation will concentrate on routes up to 300 km to 500 km where battery electric trucks are currently entering the market. This presents Hamburg with an early opportunity to position itself as a climate-neutral transport hub within the timeframe of its 2040 objective.

All zero-emission technologies require new infrastructure and investment. Now is the time for stakeholders to lay the ground, given that achieving the 2040 climate neutrality objective will require 10 to 20 years of infrastructure development. Uncertainty is amplified for long-distance freight transport since different jurisdictions could make diverging choices, making it essential to coordinate infrastructure planning along major routes out of ports into the hinterland well ahead of final technological choices.

A 2021 ITF study on zero carbon supply chains in Hamburg recommended that the HPA play a more proactive role in driving the decarbonisation of transporters by leveraging its position as a key transport node (ITF, 2021[38]). The Chamber of Commerce is well placed to encourage hinterland transport and logistics businesses to develop targets in line with the 2040 neutrality objective and support them in developing relevant strategies. This should include setting up networks and thematic committees within the Chamber so that all stakeholders coordinate and share knowledge as well as resources.

The challenge of reaching zero emissions over global supply chains is daunting: encouraging route-specific cooperation among transporters and jurisdictions makes it more manageable. Such efforts can reduce uncertainty and mobilise infrastructure investment, giving Hamburg a head start. This could involve promoting hinterland freight transport "green corridors", similar to the zero-emission maritime green corridor route agreed in 2022 between Hamburg and Halifax (Port of Halifax, 2023[39]).

Businesses could also benefit from immediate practical help in choosing zero-emission solutions for trips using tools such as the Port of Barcelona's "eCOcalculator" to evaluate CO_2 emissions generated by transporting a container into the hinterland (Port of Barcelona, 2019[40]). This service could be extended to in-depth analysis, assessing more complex transport chains and exploring emission reduction solutions.

Hamburg is, like many other ports, deeply embedded in a metropolitan area with a densely populated hinterland. While zero-emission vehicles will eliminate CO_2 emissions and tailpipe pollutants, and halve traffic noise, other impacts such as braking emissions and traffic congestion will remain and indeed grow along with the projected rise in freight transport. The future importance of BEVs also points to the need to associate decarbonisation efforts with energy and transport demand management.

Experience shows that businesses benefit from information exchange and networking to make transformations and reach financial decisions - they help find cost-effective solutions and lower transaction costs. Generally, business-based energy-intensive solutions are more readily considered than changing habits or processes to reduce energy inputs, even though energy savings are in fact the first "zero-emission fuel" (IEA, 2021[41]) and are of critical importance in view of the difficulties to expand renewable energy production to the required scale. Coordinated logistics harness such savings.

The ITF Hamburg case study also recommended stronger involvement of the city administration in zero carbon freight by initiating a coordination group for port cities in Europe and across the world to develop common policies. Coordination with jurisdictions that could be considered freight transport competitors is particularly relevant for the Chamber of Commerce to avoid carbon leakage. The city and its port are ideal backdrops to test and demonstrate zero-emission solutions. Effective policy advocacy requires a technological and policy watch of the issues involved should be formally established as soon as possible, so that relevant information can be shared in a timely fashion.

Actions for Businesses

The main obstacle for shippers is the lack of zero-emission transport solutions offered by transport operators. The large share of rail in freight transport out of the port of Hamburg provides carriers and forwarders with a head start in working towards zero-emission supply chains, compared with many other large ports. At the moment, the only zero-emission freight transport chain possible would combine rail transport with last-mile transport by BEVs, though in practice there are still unavoidable CO_2 emissions because road transport has not been entirely decarbonised.

Businesses can accelerate the climate neutrality transition by investing early in zero-emission solutions, even if they initially come at a "green" cost premium which some customers are willing to pay. To succeed in these new markets, companies should design market strategies that target green segments of their business.

Many stakeholders in freight transport are interrelated. Strategic partnerships between shippers, logistics providers and their transport partners can provide a collaborative environment supportive of supply chain decarbonisation. Long-term shipping contracts that include climate neutrality criteria in contract bidding can also provide an opportunity to decarbonise transport and logistics operations.

Transport and logistics businesses with large fleets could pilot zero-emission truck adoption in clusters or along transport lanes, as a first step towards decarbonisation. Networking could help deploy private depot chargers which will be central to fleet charging needs, with access encouraged through collaboration between fleet owners, truck manufacturers, utilities and infrastructure providers.

Financing and leasing companies should turn their attention to innovative business models to overcome the upfront costs associated with zero-emission trucking, particularly for smaller operators. Solutions such as trucking-as-a-service and charging-as-a-service can facilitate the transition to zero-emission trucking for fleet owners, as they are less capital-intensive (ICCT and ECTA, 2022[42]).

The larger businesses have dedicated research and innovation centres focused on preparing for the logistics of the next decade and beyond. Hamburg, as a major industrial and transport hub, has the critical mass of knowledge and resources to support these efforts. The Chamber could work towards linking small businesses with these efforts and combining them with local scientific advice.

Key actions

Immediately

- The Hamburg Chamber of Commerce could prepare hinterland freight transport green corridors through cooperation among transporters, vehicle manufacturers and infrastructure providers.
- The Hamburg Chamber of Commerce could provide "ecocalculator" tools to evaluate CO_2 emissions generated by transporting goods, considering emission options for individual trips and routes.
- The Hamburg Chamber of Commerce could establish a technological and policy watch for monitoring and sharing information related to zero-emission options in freight transport and logistics.
- Shippers should introduce decarbonisation and energy efficiency criteria in bidding processes for shipping contracts to incentivize carriers to adopt greener practices.
- Transport and logistics businesses should pilot the adoption of zero-emission trucks in clusters or specific transport lanes as a step toward full decarbonisation.

- Transport and logistics businesses should design a green/climate-neutral service proposition, identify target markets and create a pricing strategy for zero-emission services.

By 2030

- The Hamburg Chamber of Commerce could promote leveraging Hamburg's position as a clean energy hub, taking into account the shift from transition fuels to decarbonisation.
- The Hamburg Chamber of Commerce could promote coordinated energy-saving logistics by providing information and networking.
- Shippers should form strategic partnerships with logistics providers and transport partners to collaboratively work toward supply chain decarbonisation.
- Transport and logistics businesses should collaborate to deploy private depot chargers for electrified heavy-goods vehicles, especially where public infrastructure is lacking.

By 2040

- The Hamburg Chamber of Commerce could monitor and coordinate with other jurisdictions, notably on railway renovation and other infrastructure improvements.
- The Hamburg Chamber of Commerce could continue taking account of the urban environmental impacts of freight transport, including PM emissions, traffic congestion and noise.

The transition to climate neutrality in key manufacturing sectors

As discussed in Chapter 1, manufacturing employs more than 100 thousand people in Hamburg and generates around 12% of value added. It is the most productive economic sector in the Hamburg economy. Some manufacturing sectors will be particularly hard to make climate-neutral (OECD, 2023[43]), mostly those producing basic materials including basic metals and oil refining. They employ around 8000 workers in Hamburg. In addition, cement is produced close to Hamburg.

These sectors have depended particularly strongly on fossil fuels as inputs, both as energy carriers and as raw materials. They require high temperatures in production processes which do not easily lend themselves to electrification. Long-lived fixed capital assets characterize production in the sectors. Replacement of existing equipment may therefore need to be net-zero consistent starting in 2025 even to reach climate neutrality in 2050 (Material Economics, 2019[44]).

The manufacturing activities producing basic materials are located within the port. The port helps better integrate production in global value chains. Indeed, the production of basic materials permeates the value chains of a wide range of manufacturing products. It also depends on imports of materials and energy inputs. For example, in the EU the manufacture of basic iron and steel alone employs 375.000 workers. Upstream and downstream jobs could be 5 times as many (Oxford Economics, 2019[45]). Aluminium production employs about 230.000 workers and about 1 million workers when including indirect employment (European Aluminium, 2015[46]). Successful decarbonisation of basic materials production is therefore important beyond Hamburg on a European scale.

Transformations towards the circular economy are particularly important in manufacturing industries producing basic materials. Reducing demand for raw materials saves energy and other resources and avoids process emissions. (Sun, Lettow and Neuhoff, 2021[47]). Circular economy approaches, for example, could reduce CO_2 emissions in plastics, steel, aluminium, and cement by 56% in developed economies by 2050 relative to a baseline with no further climate action and no major shift in materials intensity or industry structure (Johnson et al., 2021[48]; Sharmina et al., 2021[49]; Material Economics, 2019[44]).

Beyond climate, raw materials extraction and processing also account for substantial water, soil and air pollution and ecosystem destruction worldwide. These environmental impacts are particularly severe in basic metals production, notably steel and copper. Most global environmental impacts of extraction and processing of these key materials are projected to at least double between 2017 and 2060 if recent materials use and policy trends continue (OECD, 2019[50]). These trends need to be reversed to address the global interrelated challenges from climate change, biodiversity loss and the degradation of land.

The circular economy requires exchanging and reprocessing materials or shared-use assets among manufacturing plants or their customers. It also requires moving towards producing goods that can be used for longer, be reused, or be available for shared use. Reducing transaction costs is therefore important, for example, to ensure the precise composition of materials or the repairability of components is known. Digitalisation and industrial symbiosis can contribute:

- Digitalisation can support the circular economy practices in the industry through improved tracking of product and materials composition. Major opportunities include marking technologies, low-cost sensors, and real-time tracking to provide better information on materials composition as well as automation, for example in sorting (Material Economics, 2018[51]). Other opportunities are geolocation technologies to indicate asset locations or blockchain to store information (OECD, 2020[52]). Digital technologies also reduce transaction costs in innovative circular economy business models, such as in the provision of capital goods as a service (Barteková and Börkey, 2022[53]).
- Industrial symbiosis, or closed-loop recycling as it is sometimes called, involves the use of by-products from one firm as inputs for another. Industrial symbiosis reduces intermediaries and is most common in industries that produce pure and homogeneous materials, such as the chemicals industry. Some of these relationships may develop organically or are the result of carefully planned industrial parks (OECD, 2019[54]). The partnership of industrial establishments across sectors, sharing infrastructures and their material inputs and outputs (including waste) can also optimise resource use.

Decarbonising manufacturing is a particular challenge in view of global competition and different climate policy ambitions across countries. In principle, the pricing of environmental footprints in value chains could address this challenge. The EU carbon border adjustment for steel, aluminium and cement products addresses this challenge to some extent but may expose domestic producers to deteriorated competitiveness in third markets, where such rules do not exist. Broad international alliances for the taxation of environmental footprints would therefore be useful.

The remainder of this section will take a closer look at manufacturing activities in Hamburg that are difficult to decarbonise, in particular, steel, copper, aluminium and oil refining.

Steel manufacturing

Steel is essential in the global economy for its wide manufacturing and construction applications such as automobiles, industrial machinery, buildings, railways, and bridges. Steel is also vital for capital goods and infrastructure in the net-zero transition, such as for electric vehicles and wind power. The demand may overall remain stable or slightly increase in Europe, while global demand may increase by more than a third (IEA, 2020[55]). Among heavy industries producing basic materials, iron and steel production is particularly dependent on freight services, relative to value-added, alongside non-metallic minerals (OECD, 2023[43]).

Hamburg has three installations in steel manufacturing in the EU ETS, all owned by ArcelorMittal. In 2019, emissions of the three installations summed up to 346 thousand $MtCO_2$. ArcelorMittal is one of the world's leading integrated steel and mining companies. Globally, they produce 69.1 million tonnes of crude steel, of which 0.9 million tonnes in Hamburg. In Hamburg, the company also produces billet and high-quality wire rods, and its production is mainly sold to automotive and engineering customers in the European market.

In the long term, decarbonisation will require the adoption of new technologies for steel production on the basis of mined raw iron (primary production). The main option without relying on Carbon Capture and Storage (CCS) is hydrogen-based direct reduction with "green" hydrogen (H-DR) (IEA, 2020[55]; Material Economics, 2019[44]). The H-DR route for making steel may be more attractive in regions with access to low-cost green hydrogen. Green hydrogen produced with renewable electricity replaces coal or natural gas as the reducing agent. Steel production in Hamburg could therefore benefit from a hydrogen hub in Hamburg to obtain access to hydrogen at the lowest possible cost, an issue taken up below. Hydrogen needs can also be avoided by outsourcing the production of iron pellets to locations with cheap hydrogen production. Iron pellets are easily transported.

Currently, most steel worldwide is produced with a combination of blast furnaces and basic oxygen furnaces (BF-BOF). Another method to produce steel is through direct reduction and electric arc furnaces (DRI-EAF). DRI-EAFs have the potential to be fully decarbonised, in terms of Scope 1 and Scope 2 emissions, using green hydrogen and climate-neutral electricity sources (Wang et al., 2021[56]; Bataille, 2020[57]) without large-scale replacement of production equipment (Box 3.5).

Steel in Hamburg is produced using Europe's only DRI-EAF furnace (Eurofer, 2022[58]). Hamburg steel production is therefore relatively well-placed for decarbonisation. Even so, the new production processes will have wide implications for infrastructure such as raw materials storage and processing, transport and supply of energy, power and steam distribution and generation (Material Economics, 2019[44]). As elsewhere, steel manufacturing in Hamburg is particularly energy-intensive.

ArcelorMittal is preparing to make the switch to hydrogen instead of natural gas in the iron ore reduction process. A project is underway to test hydrogen DRI for iron production on an industrial scale, integrating it into the EAF steelmaking process (Reuters, 2021[59]). The objective is to reach industrial commercial maturity of the technology by 2025, initially producing 100,000 tonnes of climate-neutral iron a year. The German government has expressed its intention to provide €55 million of funding support towards the plant's construction (Box 3.6).

Box 3.5. Steel production methods worldwide

The ironmaking and steelmaking phases are the most emissions- and energy-intensive phases in the making of steel products minerals. Currently, most steel worldwide is produced with a combination of blast furnaces and basic oxygen furnaces (BF-BOF). Another method to produce steel is through direct reduction and electric arc furnaces (DRI-EAF).

- **BF-BOF:** Iron ores are heated with highly purified coal in a blast furnace (BF). The carbon in the coal binds with the oxygen of the iron ore as CO and CO_2, leaving the elementary iron in a melted state. Then, various other elements, such as carbon, chromium, zinc and nickel, are added in a subsequent alloying process in a separate basic oxygen furnace (BOF) to produce steel.
- **DRI-EAF:** In direct reduction (DR), the iron ore reacts with hydrogen (H_2) combined with carbon monoxide (CO) in syngas, now usually made using methane (mostly from natural gas) and water. The elementary iron is then melted in an electric arc furnace (EAF) and alloyed as necessary to produce steel (Wang et al., 2021[56]; Bataille, 2020[57]).

Steel products can be made from iron ore or scraps, or a combination of both. Recycled secondary steel is usually made using the EAF method. On average, the BF-BOF method emits 2.3 tonnes of CO_2e per tonne of steel produced. Natural gas-driven DRI-EAF plants emit 0.7 tonnes of CO_2 per tonne of steel produced (Wang et al., 2021[56]; Bataille, 2020[57]).

Source: (OECD, 2023[43])

Production costs and investment

Based on current estimates of the levelised costs of production for commercial-scale plants, producing one tonne of carbon-free primary steel (i.e. without using recycled materials as inputs) is at least 8-9% more expensive than today's main commercial production routes. However, this is based on CCS-equipped DR, with hydrogen produced from natural gas, and innovative smelting reduction processes. Deploying CCS at scale is subject to major uncertainty. All other options for decarbonising steel, except for recycling, cost at least 30% more. The hydrogen-based DR route typically raises costs by around 35-70% for bulk steel (IEA, 2020[60]), with energy efficiency and electricity price being determining factors. The costs can also vary depending on access to green hydrogen. In downstream products, metal parts may perhaps cost 10% more and would only add $300-400 to a car for example (0.5-2%) (Bataille, 2020[57]). Falling costs of green hydrogen production will lower this gap.

The viability of emission-free production of primary iron and steel in Hamburg will depend on electricity and hydrogen prices. A more granular distribution of electricity prices over time and space would result in more efficient energy markets and would give energy-intensive industries in Northern Germany, including Hamburg, a better chance. The overall regional balance of energy supply and demand in Hamburg is therefore also likely to influence prices at which electricity will be delivered to industry.

Integrating circular economy practices in steel and other basic metals production is key to lowering Scope 3 emissions. While ArcelorMittal does not publish Scope 3 emissions, emissions from copper production in Hamburg suggest they can be a high multiple of Scope 1 and 2 emissions. Beyond climate, the extraction of raw iron and its processing are among the most environmentally damaging materials processing activities, with a broad range of strong, adverse impacts on energy use, climate, human toxicity as well as terrestrial and ecosystem degradation, that would be set to rise further on current world-wide materials use trends (OECD, 2019[50]). Integrating circular economy practices is therefore of strategic interest especially in steel.

The use of steel scrap mainly uses electricity and therefore entails much lower GHG emissions (Wang et al., 2021[56]). It can be fully decarbonised without resorting to hydrogen and with substantially lower energy input. Expanding steel production on the basis of scrap processing can also be an interesting option if electricity or hydrogen costs are locally too high. There are at present limitations to relying on scraps for recycling (IEA, 2020[61]) both in terms of availability of scraps and quality hence, mitigation strategies for steel cannot rely solely on recycling (Material Economics, 2019[44]). In particular, copper contamination in steel scrap limits high-value recycling.

Box 3.6. Steel companies with net zero targets and EU initiatives

In the steel industry, there are industry-wide and firm-specific initiatives to reach climate neutrality. Several EU steel companies have plans to start piloting H-DR, including Salzgitter, SSAB, ThyssenKrupp, and Voestalpine (Material Economics, 2019[44]). Some companies that have net zero emissions ambitions include ArcelorMittal, Thyssenkrupp, Acerinox and Posco. Companies mention technology levers such as green hydrogen, direct reduction and renewable electricity. Also, the use of CCUS is included to keep current assets running, and thus reduce stranded asset risk, and make more short-term progress. To reach these goals, companies highlight the importance of the provision of affordable clean energy, circular economy infrastructure, sustainable finance for developing climate-neutral steelmaking and a global level playing field. (Acerinox, 2021[62]; ArcelorMittal, 2021[63]; Posco, 2021[64]; Thyssenkrupp, 2020[65])

The EU Ultra Low Carbon Steel program (ULCOS) is a cooperative R&D initiative on CO_2 emissions reduction solutions in steel production between the European Commission, major EU steel companies and other research partners (Quader et al., 2016[66]). One technology that was explored is the Hisarna process, which produces a more concentrated CO_2 waste gas than BF-BOFs, facilitating carbon capture. The Hisarna process could also be installed next to existing steel facilities but would need to be used with carbon capture and storage (CCS). It would not disrupt existing supply chains and labour forces (Bataille, 2020[57]).

In Sweden, a Hydrogen Breakthrough Ironmaking Technology (HYBRIT) project was launched in 2016 in cooperation with SSAB. The Swedish HYBRIT hydrogen DR-EAF project uses hydrogen, produced by electrolysis using fossil-free electricity, to produce primary iron ore pellets. Electricity prices are a main production cost parameter (Pei et al., 2020[67]). The project delivered its first carbon-free steel to Volvo in August 2021 (SSAB, 2021[68]). Although volumes are still small, SSAB plans to lift production to an industrial scale by 2026. The company also plans to convert a production plant in Sweden by 2025.

In Germany, ArcelorMittal supports this ambition by operating a hydrogen DR-AEF facility in Hamburg. With partial funding from the German government, it is building a commercial-scale demonstration plant that will use either green hydrogen and carbon monoxide or pure green hydrogen for direct reduction (ArcelorMittal, 2021[69]).

Other projects to produce green steel using pure green hydrogen instead of natural gas currently being developed in Europe include GrInHy and SALCOS-MACOR by Salzgitter in Germany and SuSteel and H2Future by Voestalpine in Austria. ThyssenKrupp is planning on eventually converting its Duisburg plant In Germany to produce steel with hydrogen.

One promising alternative to hydrogen is using electricity to reduce iron ore through electrolysis. This method is being explored by Boston Metal in Massachusetts, and Luxembourg-based Arcelor Mittal (Fennell et al., 2022[70]).

Source: (OECD, 2023[43])

Copper manufacturing

Copper is an indispensable metal for decarbonisation. With its high electrical conductivity and corrosion resistance, it plays a significant role in electrification, the deployment of renewable energy and electric car production. The global copper demand could, on account of these demands and on account of income growth, triple in 2050 compared to the 2010 level (Elshkaki et al., 2016[71]).

70% of the total GHG emissions from the entire copper production value chain occur in mining sites, 23% are from smelting and refining, an activity carried out in Hamburg, and 7% in transport and the recycling of sold products (International Copper Association, 2023[72]). This is reflected in high scope 3 emissions of primary copper production in Hamburg (Box 3.7). Other negative impacts of copper mining and processing on the environment include water pollution, soil contamination and ecosystem degradation. While less wide-ranging than for iron processing, these impacts could increase particularly strongly for copper by 2060 on recent policy and economic development trends (OECD, 2019[50]).

Circular economy practices can contribute substantially to reducing Scope 3 emissions and other environmental impacts along the value chain, especially from copper mining. Copper production from recycled materials (secondary production) involves only 35% of the carbon footprint production that results from the use of raw materials from copper mines ("primary production") (Grimes, Donaldson and Grimes, 2015[73]). The total environmental impact of secondary copper is only 1/8 of the primary copper production process. This entails initiatives to advance sorting methods, enhance fabrication yields, prolong the lifespans of products, and more efficient use of the existing copper stock (Watari et al., 2022[74]). At present, 50% of copper products are recycled, and 30% of global copper demand is met by recycled inputs (OECD, 2019[50]). Even 100% recycling rates will not be able to end the need for primary supply by 2050 (Hund et al., 2023[75]).

Energy intensity is high in both primary and secondary production. A mix of electrification, green hydrogen and ammonia, produced from green hydrogen, as well as the reuse of waste heat can decarbonise the energy use. The energy density of ammonia, produced emission-free from green hydrogen, is higher than for hydrogen itself (Valera-Medina et al., 2018[76]) and therefore serves for achieving elevated process temperatures (The Royal Society, 2021[77]). The planned ammonia terminal may enhance the competitive edge of business in Hamburg. Electricity prices will also be an important determinant of competitiveness.

As in the case of steel, process CO2 emissions result from the use of natural gas in the reduction process. Hydrogen can serve as a substitute for natural gas as a reducing agent in both primary and secondary production. Hydrogen can remove oxygen from the copper mineral melt, generating water vapour instead of CO2 (Luidold and Antrekowitsch, 2007[78]). GHG emissions also come from the separation of organic chemicals from secondary copper materials (Chen et al., 2019[79]). These processes entail the intricate separation of over 50 materials, each with distinct thermodynamic and metallurgical characteristics, from a single product (IEA, 2021[80]). Carbon Capture and Storage (CCS) can help to decarbonise the secondary production where CO2 is generated when separating copper from other materials, such as plastics.

Box 3.7. The decarbonisation strategy of Aurubis AG in Hamburg

Aurubis, a metal manufacturing company headquartered in Hamburg, specialises in producing copper cathodes and related other copper products in several production sites. In 2021, Aurubis reported 559,000 tonnes of Scope 1 emissions, 1,047,000 tonnes of Scope 2 emissions, and 6,181 000 tonnes of Scope 3 emissions across their business sites. Out of their Scope 1 emissions, 165 000 tonnes originated from the Hamburg plants. The Hamburg site of Aurubis is taking the lead in the group on decarbonisation and carries out research and development for the group.

Aurubis is working towards decarbonising its value chain and aims to reduce Scope 1 and 2 emissions by 50% by 2030. It is expanding its long-term power purchase agreements with energy companies that have a large share of renewable energy in their portfolios. Additionally, Aurubis generates its own electricity through on-site solar plants and utilises waste heat to generate electricity. Waste heat also plays a role in supplying heat and processing steam as well as feeding district heating for buildings in Hamburg.

Supported by the Federal government's hydrogen supply strategy, Aurubis has conducted trials replacing natural gas with hydrogen as a reducing agent in anode furnaces. The use of ammonia in rod plants is also being explored as part of a collaborative effort on hydrogen between Germany and the United Arab Emirates (UAE). In the initial stages of the production process, an ammonia co-firing trial is set to replace 20% of the natural gas fuel for six to eight weeks this year.

To reduce Scope 3 emissions, Aurubis aims to achieve a recycled content share. Its Hamburg facility specialises in recycling raw materials. In 2022, it decided to invest € 190 million to enhance recycling in the Hamburg recycling operations. Aurubis closing material loops through customer collaborations and product marketing. The company offers tailored solutions for customers to recycle copper and other metals. This includes selling copper scrap back to Aurubis and receiving refined copper in exchange.

Source: (Aurubis AG, 2023[81]); (Carbon Disclosure Project, 2022[82])

Decarbonising aluminium

Demand for aluminium may increase as it can displace other materials to reduce weight and thereby energy use (Pedneault et al., 2021[83]). For example, replacing steel with aluminium reduces energy consumption in cars, a key consideration for more sustainable car production. In aluminium production primary production from mined raw materials has on average a greenhouse gas emissions footprint twenty times higher than secondary production from scrap. Secondary aluminium production also boasts a lower environmental footprint more generally and much lower energy use. (Liu and Müller, 2012[84]; Liu, Bangs and Müller, 2013[85]). Primary and secondary aluminium cannot be substituted with the given equipment. The Hamburg production of aluminium consists of primary aluminium only.

Secondary aluminium production represents over half of world aluminium production. Aluminium is quasi-infinitely recyclable. However, scrap availability still limits recycling, reflecting long product life and the short history of mass aluminium production (IAI, 2021[86]). Circular economy strategies also need to overcome alloy inter-contamination in recycling and to further reduce transformation and fabrication losses (Pedneault et al., 2021[83])

Primary aluminium producers are large electricity consumers (IEA, 2020[61]). The price of electricity is hence a major production cost determinant. The Hamburg aluminium production plant reports it can respond flexibly to variable renewable energy supply. Aluminium production would therefore benefit from the granularity of electricity pricing, so its positive role in system stability can be valued. Even so, aluminium production also requires some baseload electricity to ensure continuity of production, which offshore wind could supply. The extent to which electricity prices need to incorporate network costs may be key. Reflecting high electricity needs and the significant share of coal in German electricity Scope 2 emissions are about 3 times higher than Scope 1 emissions in Hamburg aluminium production. No information on Scope 3 emissions is available.

Climate-neutral technologies, in particular inert anode smelting, need to reach significant market penetration in the next two decades (Liu, Bangs and Müller, 2013[85]) and are not yet deployed in Hamburg. Replacing carbon anodes with inert anodes eliminates direct emissions, including process emissions (IAI, 2021[86]). Following this route makes CCS redundant. The electricity requirement for inert anode smelting may be 50% higher than for the current smelting process (Pedneault et al., 2021[83]), although this may come down with technological developments. Moreover, stranded asset risks are substantial as it is currently unclear whether it is feasible to retrofit existing installations, which have an economic life of at least 20 years (Pedneault et al., 2021[83]).

Employment effects of decarbonising basic metals

Employment in the manufacture of basic metals, including aluminium, copper and steel, in the EU-27 may slightly increase until 2030, according to Cedefop's European Green Deal (EGD) scenario (Cedefop, 2021[87]). The OECD ENV Linkages model finds that the effect of decarbonisation policies will cause a slight decrease in employment in aluminium and a slight increase in steel by 2040 across OECD and most EU countries, though the large EU countries may lose some employment.

Decarbonised steel production may have significant employment potential, particularly for science, technology, engineering and mathematics graduates (University of Cambridge Institute for Sustainability Leadership (CISL), 2020[88]). As part of the shift towards the circular economy, workers will be expected to possess strong digital skills as well as an understanding of environmental management and knowledge to support circular business models (University of Cambridge Institute for Sustainability Leadership (CISL), 2020[88]). A major challenge for the sector is the ageing workforce which may slow skills renewal and green skills.

Oil refining

Refined petroleum products have two uses: energy use (fuels) and non-energy use (raw material or "feedstock" for petrochemical products). Direct emissions from the manufacture of refined petroleum products are among the highest, relative to value-added, across manufacturing sectors, but the Scope 3 emissions, notably from using final oil products, are substantially higher than direct emissions (Jing et al., 2020[89]). With the electrification of road transport, global demand for fuel oil products will decrease strongly (IEA, 2021[90]).

Where electricity cannot substitute energy provided by fossil fuels, new fuels, including biofuels, hydrogen and synthetic fuels will substitute crude oil. Of these, only biofuel production can be based on existing oil-refining production assets. Oil refineries converted to biofuel plants can reduce the stranding of assets and reduce capital costs of biofuel production (Karka, Johnsson and Papadokonstantakis, 2021[91]). Other benefits include using shared infrastructures, such as for transport, and logistic networks from the oil refining infrastructure.

Demand for oil products as "feedstock", such as to produce plastics, may by contrast increase at a global scale for some time. Overall, the IEA (2021[90]) expects a decrease in global oil demand by around 70% by 2050 in a net-zero-emissions scenario but still expects some oil to be produced in 2050 at a global level. Even so, meeting the EU climate neutrality target may reduce fossil fuel use substantially even as feedstock by 2050, especially if CCS is avoided. This could also concern petrochemical production in Hamburg (Box 3.8).

One possible replacement is with biofuels. Green hydrogen combined with carbon capture and use (CCU) to produce synthetic hydrocarbon fuels as well as chemical plastics recycling can be alternatives but are energy-intensive. Another possible replacement is biofuel. Biofuel is likely to face high demand, for feedstock and fuel use. Its production will compete with other land uses, notably for food production, biodiversity protection and carbon sinks. Its sustainable sourcing is necessary for climate neutrality. The demands on biomass therefore risk exceeding sustainable supply substantially. Sustainable biofuels will have unavoidable priority uses. These include aviation and shipping fuels, as well as feedstock to produce petrochemicals and plastics or fibre-based substitutes for plastic products (Material Economics, 2019[44]).

According to the EU ETS, there are six oil refining installations in Hamburg. The bulk of emissions arise in the production of diesel transport fuels which do not serve priority use. Unless this production can be repurposed to produce biofuels, the production is likely to cease in the transition to climate neutrality. Biofuel processing is thus far particularly strong in countries with high biofuel production potential, such as in Brazil or Sweden. This suggests biofuel processing may not play an important role in Hamburg. On the other hand, Hamburg's economy depends on sectors where biofuel use will be a priority, notably in transport, as well as in the production of some petrochemical products (Box 3.8).

Box 3.8. Decarbonising oil refining: The example of Hywax in Hamburg

Hywax in Hamburg produces waxes, petroleum jellies and wax emulsions for different types of products, such as insulation materials, cosmetics, coated paper, adhesives for cardboard, food products, etc. The raw materials used are to a large extent by-products from crude oil refining and more specifically from the production of base oils, which are for example used for motor oils. Hywax in Hamburg is one of the leading companies in wax production in Europe. It employs around 350 people and is an important activity for the port. It loads about one ship (barges and larger vessels) per day with waxes.

Hywax used to produce their finished products 99% from raw materials derived from crude oil. Nowadays it uses 70-80% of crude-oil-derived raw materials and the rest are natural and synthetic waxes. Waxes are then kept in tanks at high temperatures (60-80°C) to remain liquid.

Hywax aims to be **climate-neutral by 2040** through three main measures:

1. **Optimisation:** the company plans to replace older tanks and pipes with newer, better-isolated ones to reduce heat loss. It also plans to invest in newer, more energy-efficient machines.
2. **Electrification:** the company plans to switch from natural gas (which is mainly used in a co-generation unit to produce steam and electricity) to a) heat pumps to provide heat for wax production and storage, and b) external sourcing of renewable electricity as the main energy source.
3. **Harnessing variable renewables:** As waxes only lose around 1°C in temperature per day if the tank heating is interrupted, the variability of renewable energy sources is not an issue but an opportunity. Waxes in the storage tanks can be heated up to 95°C during periods of high availability of renewables. Wax production is thus well placed to benefit from low-cost intermittent renewables supply.

Nevertheless, for fully climate-neutral wax production, feedstocks need to be further replaced by natural-based and synthetic waxes. For synthetic waxes, the options are 'green' hydrogen combined with carbon capture and use (CCU). The processing is similar to the production of synthetic sustainable aviation fuel (SSAF).

Source: (OECD/Hywax, 2023[92])

Employment effects of decarbonising oil refining

A substantial reduction in this employment can be expected. According to Cedefop's European green deal (EGD) scenario, direct employment in manufactured fuels in the EU-27 will decrease by about 10% until 2030 (Cedefop, 2021[87]). This is additional to an already expected decline of 4% in this sector in Cedefop's baseline scenario. A substantially bigger decline is expected beyond 2030, as activity needs to be phased out.

A study on the transferability of skills in the UK offshore energy workforce found that soft skills, business skills and other non-technical skills in the oil sector tend to be highly transferable to adjacent clean energy sectors such as offshore wind, carbon capture and storage, hydrogen or other industrial sectors (de Leeuw and Kim, 2021[93]). However, transition training and upskilling will be required.

Key actions

Immediate action

- Businesses should replace long-lived capital assets in a way that is consistent with climate neutrality.
- The Chamber of Commerce could seek to support broad international alliances for the taxation of environmental footprints in basic materials production, extending them beyond the EU.
- Scope 3 emission accounting should be provided and included in climate neutrality targets and action plans of all Hamburg basic material producers.
- The HCC could assess, with key businesses, employment and worker skills implications from the transformations in manufacturing, in particular in oil refining, and identify actions to maintain attractive job prospects for affected workers.

By 2030

- Key businesses in basic materials manufacturing should define circular economy strategies for the decarbonisation of value chains.
- The Chamber of Commerce and key businesses in basic materials manufacturing could assess the key determinants of making climate-neutral production viable in Hamburg, including the cost of hydrogen and electricity.

By 2040

- Key businesses in basic materials manufacturing should largely eliminate direct missions in Hamburg production sites, with the remainder to be offset with the purchase of carbon credits.

The potential for a green hydrogen hub in Hamburg is large

Green hydrogen is an essential tool for reducing emissions in all sectors that are difficult to electrify (IRENA, 2020[94]) (IEA, 2023[95]) (IEA, 2022[96]). Several of these are present in Hamburg, including steel and copper production. Hydrogen-based fuels for shipping are also particularly relevant for Hamburfg The European Union's hydrogen strategy (European Commission, 2022[97]), as well as RePowerEU (European Commission, 2022[98]) place green hydrogen as a major instrument to lower CO2 emissions, as well as the German National Hydrogen Strategy and the Hydrogen Strategy for Northern Germany (Federal Ministry for Economic Affairs and Energy, 2020[99]).

As a key European industrial port, Hamburg holds a comparative advantage. The IEA emphasizes the pivotal role of "*mak(ing) industrial ports the nerve centers for scaling up the use of clean hydrogen*" (IEA, 2019[100]). Hamburg's strategic plan involves establishing itself as a prominent green hydrogen hub in the North Sea region (Box 3.9), with multiple industries in the area already demonstrating a strong inclination toward procuring green hydrogen (Future Hamburg, 2022[101]).

Box 3.9. The Hamburg green hydrogen hub

German green initiative for hydrogen

Initiated by the Northern German Chamber of Industry and Commerce, the green initiative for hydrogen gathers Hamburg, along with Bremen, Schleswig-Holstein, Mecklenburg Western Pomerania and Lower Saxony, with initial subsidies of 9 billion euros. It aims at turning Northern Germany into the strongest region for green hydrogen in Europe. Nearby the port of Hamburg, the Helmholtz Centre gather scientists working on a new kind of storage facility for hydrogen.

Hamburg Hydrogen Network

The “Hamburg Hydrogen Network” (*Wasserstoffverbund Hamburg*), composed of Airbus, Arcelor Mittal, Gasnetz Hamburg, GreenPlug, Hamburger Hafen und Logistik AG, Hamburg Port Authority and the Green Hydrogen Hub (Luxcara and Wärme Hamburg) are all developing projects to use green hydrogen. For example, Airbus announced that it will commercially offer hydrogen-powered aircraft by 2035; Arcelor Mittal plans to use hydrogen to produce around 100 000 tons of sponge iron per year for steel production as of 2024 and Hamburger Hafen und Logistik AG aims to use heavy-duty vehicles with hydrogen fuel cells on its terminals and last mile container traffic.

Source: (Hamburg Invest, 2022[102]) (Hamburgnews, 2020[103]) (Future Hamburg, 2022[101]) (Future Hamburg, 2022[104]) (HHLA, 2022[105])

Hamburg can contribute to satisfying hydrogen demand beyond its borders

Global hydrogen demand may rise by around 50% by 2031 even in the IEA's Announced Pledges Scenario. Announced pledges will need to be reinforced to reach Paris Agreement climate targets, which should raise hydrogen demand further. In the industry sector the demand comes mainly from the chemical and steel industries (IEA, 2022[96]). Given the presence of steel and copper production as well as some petrochemical production within the oil refining industry in Hamburg, the demand for local green hydrogen is anticipated to be substantial. Hydrogen-based fuels hold potential for maritime shipping, including ammonia. In heavy-duty road transport, hydrogen fuel cell vehicles may be an option, especially for heavy-duty transport over the longest distances, though electric solutions appear likely to be predominant because they are lower-cost. This especially applies to Hamburg, where rail takes on most long-distance land-based transport from and to the port. Germany has initiated trials of fuel cell trains and is working on a project to incorporate hydrogen in aviation.

Meeting local hydrogen needs requires infrastructure to transport, store and process hydrogen. This infrastructure will be subject to scale economies. Unit costs may be lower with higher demand. The port's potential could therefore serve as a hub for the use of hydrogen beyond local needs. Hamburg is close to regional industrial hubs with large green hydrogen demand. Figure 3.11 illustrates the projected spatial distribution of hydrogen demand in 2050, considering the current production sites are maintained. Nearby regions in Belgium, the Netherlands and North Rhine-Westphalia, account for the highest green hydrogen demand, more than 50% of European demand, although they can also be served by the Rotterdam, Antwerp and Dunkerque ports.

Figure 3.11. Spatial distribution of hydrogen demand for decarbonising chemicals and steel production

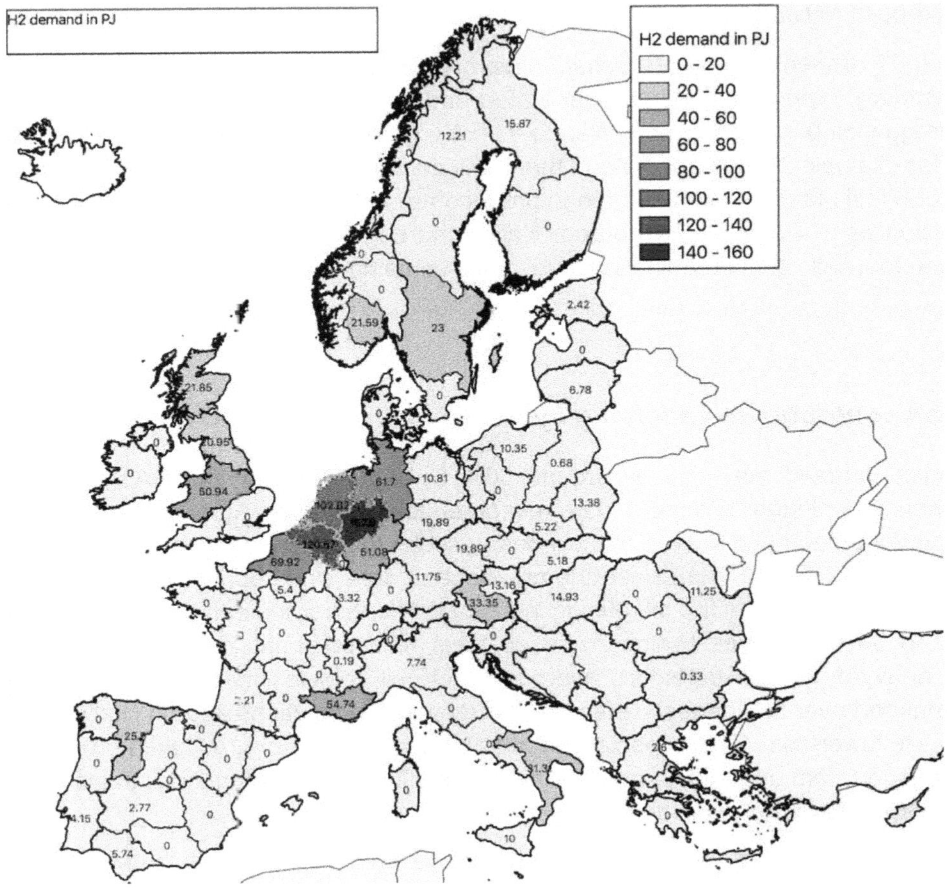

Note: assumes a development according to the "New Processes"-pathway (ENTSO-E, 2014[106])
Source: (OECD, 2023[43])

Most hydrogen will need to be imported

Germany is a leader in the development of green hydrogen production projects, but they fall short of near-term demand. The largest is the AquaVentus project in which Hamburg participates (Box 3.10). In addition, Germany has launched the Power-to-X (International PtX Hub, 2023[107]) and H2Global (H2 Global Stiftung, 2023[108]) initiatives to provide knowledge platforms which connect stakeholders from politics, business, science and civil society in order to discuss scale-up. Power-to-X helps identify market potential and funding. It organizes training and technical advice, while H2Global participates in the development of infrastructure that turns electricity into carbon-neutral fuels (PtX), market models and supply chains for green products. H2Global is a competition-based market instrument that aims at promoting a timely and effective ramp-up of the PtX market on an industrial scale.

The projected installed capacity falls short of demand (Figure 3.12) (IEA, 2022[96]); therefore, importing green hydrogen into Hamburg will be essential.

Box 3.10. Green hydrogen production projects – selected examples

AquaVentus project

The AquaVentus project will be developed in the German North Sea area to use electricity from offshore wind farms to operate electrolysers installed in the North Sea on an industrial scale. The aim is to set up electrolysis units with a total capacity of 10 GW by 2035, enough to produce 1 million metric tons of green hydrogen.

Moorburg power plant site

The site of the former coal-fired power plant Moorburg aims at decarbonising the entire port economy thanks to a 100 MW electrolyser project and the conversion of the coal power plant in the Moorburg Quarter. According to estimates, the plan can produce 2 tons of hydrogen/hour. The Moorburg site will be connected to local companies by a pipeline system to supply the port's big industrial companies: the HH-WIN (Hamburg's hydrogen-industrialism-network) will be 45 km long and will provide green hydrogen by 2030 for the steel, copper and aluminium industries. It will also be connected to the national hyperlink pipeline system.

Figure 3.12. National hydrogen targets of Germany

Sector		Target volume and time horizon
Renewable hydrogen production	Electrolysis capacity	5 GW (2030); 10 GW (2035-2040)
	Renewable hydrogen production	14 TWh/a (2030); 28 TWh/a (2035-2040)
	Renewable power consumption	20 TWh/a (2030); 40 TWh/a (2035-2040)
Overall demand	Hydrogen demand	55 TWh/a (2020); 90-110 TWh/a (2030)
	Power-based energy carrier consumption	110-380 TWh/a (2050)
	Hydrogen consumption in industry	55 TWh/a (2020); at least 65 TWh/a (2030)

Source: (Weltenergierat, 2020[109])

Producing green hydrogen requires an important amount of renewable electricity sources (solar, wind or hydropower) representing more than half of hydrogen production costs. The renewable power production potential in northern Germany, while substantially stronger than in the south, is modest even if compared to European regions (Figure 3.13). The lower renewable power production potential increases the cost, which is likely to be higher than the cost of imported hydrogen (IRENA, 2022[110]).

Beyond the potential for wind and solar power which are the highest in the Middle Eastern and African countries, the cost of imported hydrogen depends also on soft factors such as government support, and political stability. Spain, Australia, Chile, and Morocco stand out as countries with the greatest potential to emerge as net hydrogen exporting nations (IRENA, 2022[111]). Importing green hydrogen from these countries will be more economically attractive compared to local production by 2030 (Hydrogeninsight, 2023[112]).

Figure 3.13. Renewable electricity technical potential from wind and solar vary across regions

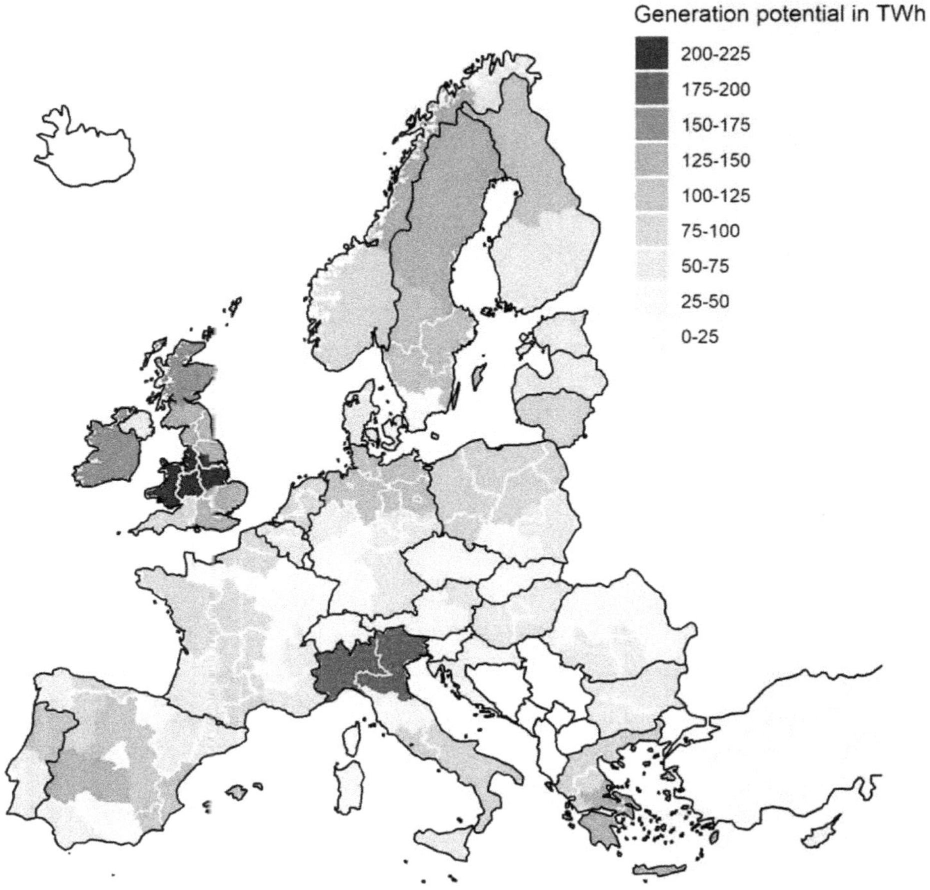

Note: Includes existing renewable electricity generation, including from other sources than wind or solar, and solar and wind technical potential
Source: (OECD, 2023[43])

Infrastructure is vital to importing green hydrogen and providing a hydrogen hub

Since production and transport distances are expected to increase to meet demand, there is a need to scale up infrastructure development to connect areas of production, import and demand.

Repurposing natural gas pipelines to transport hydrogen can cut investment costs by 50-80% compared to the building of new pipelines (IEA, 2022[96]). Hamburg could be part of a repurposed hydrogen pipeline network connected to nearby storage (Figure 3.14). However, repurposing existing natural gas pipelines may be a challenge over long distances where no multiple pipes exist. In this case, it may only be done when natural gas is out of use which could delay long-distance transport over existing pipelines. This would raise the importance of hydrogen import by ship.

Additional transport and storage capacity along the pipeline's network will be needed, as three times more volume of green hydrogen will be necessary to supply the same amount of energy as natural gas. New pipelines for ammonia (NH3, liquid) are cheaper than new pipelines for hydrogen (IEA, 2019[100]).

In May 2023, the German cabinet approved the creation of a national hydrogen core pipeline network, to become law in 2023 (Collins, 2023[113]). The network will target energy-intensive industries and connect all regions of Germany. The European Hydrogen Backbone initiative (group of 32 gas network operators) has already introduced a plan for a European-wide H2 network. Hamburg's geographical proximity to salt caverns is an advantage for integration into the network because it offers storage potential.

Figure 3.14. Repurposed pipelines can connect Hamburg to a European hydrogen network by 2040

Source: (OECD, 2023[43])

For longer distances, demand for imported hydrogen creates a need for expanding shipping and port infrastructure. Hydrogen can be transported as liquefied hydrogen (LH2) ammonia (NH3), LOHC or converted into a synthetic hydrocarbon fuel (IEA, 2022[96]). Due to the inability to transport pure hydrogen by ship, the existence of conversion and reconversion facilities is essential at both the loading and receiving terminals. By 2026, a terminal for importing green ammonia to Germany will be constructed at the port of Hamburg. It can also serve ammonia needs from future emission-free international shipping.

The Hamburg port can play a key role in importing hydrogen

Pipelines are currently the least costly way to transport hydrogen up to a distance of 2 500-3 000 km, for capacities. Hydrogen may therefore be transported to Europe via pipeline from the Middle East, through Southern Europe, as well as from Denmark and Norway. Repurposing existing gas pipelines to transport hydrogen from North Africa or the Middle East takes time, in part because natural gas will remain in use. Shipping also provides flexibility, helping to deal with any geopolitical risks. Hamburg already has existing cooperation projects with Chile, Uruguay, Argentina and Scotland, and some promising hydrogen export regions such as the United States can also export green hydrogen to Europe via shipping, making Hamburg a suitable place to deliver their products. Since shipping hydrogen is yet not developed at scale, there is a potential to invest in research, development, and deployment to lower costs. As a green hydrogen hub, Hamburg could invest to scale up technology in this field.

Shipping hydrogen gives more options on the different forms under which it will be transported. Some hydrogen-derived products, such as ammonia, organic compounds and iron pellets, offer more cost-effective transport options. Once delivered, ammonia can be converted back into hydrogen.

Hydrogen requires ambitious yet low-cost green certification

Certified green hydrogen allows industrial producers to market their products as having been produced using low or zero-carbon emission hydrogen. Certificates improve the supply chain carbon accounting (IRENA and RMI, 2023[114]). However, a cumbersome certification framework could increase costs, as some green hydrogen exporters might choose to send their product to a place with a less stringent framework. Costs could be transferred to importers of green hydrogen in Hamburg. The risk will be greater with a lack of harmonization of certification worldwide.

Currently, there are 11 voluntary or mandatory schemes worldwide. Their criteria for defining the sustainability of green hydrogen vary in scope, emissions threshold, and accounting methodology. With the growing number of certification schemes, the European Commission will probably update a list of national or voluntary internationally recognised certification schemes (Erbach and Svensson, 2023[115]), as it already does for biofuels. The CertifHy scheme, which applies only in the EU, recently submitted its RFNBO Voluntary Scheme for recognition by the European Commission (CertifHy, 2023[116]).

The IRENA Coalition for Action's ten criteria to ensure that tracking systems meet green hydrogen demand could represent an interesting basis for achieving a global system of green hydrogen certification (Box 3.11).

Box 3.11. The IRENA Coalition of Action's 10 criteria for green hydrogen tracking systems

- Develop a harmonised definition of green hydrogen
- Certify the renewable origin of the energy used to produce hydrogen thanks to technological correlation (1), geographical correlation (2), temporal correlation (3), additionality (4)
- Ensure that certificates contain sufficient information for consumers and policymakers
- Simplify the green hydrogen tracking system to avoid administrative burdens
- Implement a cost-effective tracking system
- Put in place appropriate control systems to avoid misuse or lack of transparency
- Consider interactions with existing tracking systems
- Avoid double counting
- Use taxonomy and green finance criteria to encourage compliance with green hydrogen certification requirements
- Promote international co-operation to establish globally accepted rules and requirements

Source: (IRENA coalition for action, 2021[117])

Key actions

Immediate action

- The HCC could continue to assess, with businesses concerned, as well as with regional and national government, the need for infrastructure to transport, store and process hydrogen and hydrogen-derived products, including ammonia to satisfy local needs for climate neutrality, notably in shipping, long-distance heavy-duty road transport steel production, copper production and as a feedstock in petrochemicals production.
- The HCC could assess, with businesses concerned, as well as with regional and national government, the potential for Hamburg to contribute to hydrogen demand beyond the Hamburg region, notably in North Western Europe, following up on assessments across European regions (OECD, 2023[43])
- The HCC and key businesses could quantify the local potential for green hydrogen production. They could evaluate the competitiveness of locally produced green hydrogen compared with imports.
- The Hamburg Chamber of Commerce could participate in national or international discussions to develop green hydrogen certification schemes aligned with the European Union standards

By 2030

- Increase local potential to produce renewable electricity for the supply of green hydrogen
- Scale up infrastructure development to connect areas of production, import and demand.
- The HCC could assess, with businesses concerned, as well as with regional and national government, research, development and deployment of technologies to lower costs for shipping hydrogen.

Carbon Capture and Storage in the North Sea to reduce hard-to-abate CO2 emissions from heavy industries

Carbon Capture and Storage (CCS) may be a "last step technology", that may be useful for a few hard-to-abate heavy industries (IPCC, 2022[118]; IEA, 2020[60]). One approach is to limit CCS to process emissions in industrial production, as climate-neutral options for energy-related emissions are available, as favoured in the European Union (OECD, 2023[43]). In situations where emission-free technologies are not readily available, CCS becomes a necessary requirement to achieve Net Zero Targets (NZT). Consequently, CCS is most likely to be necessary in cement production and, perhaps, in copper production, both present in the Hamburg metropolitan area. modest.

The North Sea offers major advantages for CCS storage. Indeed, several projects, mostly small-scale, have been launched around the North Sea or are planned (Box 3.12). However, a major potential for CCS development in the Hamburg region appears unlikely. As outlined below costs are high, including those for transportation. Moreover, large-scale transportation of CO2 may not require Hamburg's port infrastructure. Even so, the availability of space or transhipment facilities and compatibility with other transformation developments (such as hydrogen imports) would need to be examined.

Box 3.12. CCS projects of European countries bordering the North Sea

- In Norway, the Northern Light project is one of the most important CCS projects in Europe. The Longship project, the transport and storage component of Northern Light, is the first trans-European hub opened to all European emitters. Transport and storage capacity is up to 5 million (Mt/year) CO2, for a total storage capacity estimated around 100 Mt.
- In the Netherlands, the Porthos project is being developed in the Rotterdam port. This joint project between the Port of Rotterdam Authority, EBN and Gasunie will allow various companies to supply their CO2 emissions to a collective pipeline running through the Rotterdam port area. Only a very limited share of CO2 emissions will be captured and stored in the North Sea. Indeed, industries around the port emit around 25 Mt/year CO2 (45 chemical companies and 5 refineries), and the Porthos' project initial capacities will be limited to 2 Mt/year CO2, with a possible increase to 5 Mt/year CO2 by 2030.
- In Denmark, the Greensand project is led by a consortium of Danish and international companies, research institutes, universities and start-ups. It aims at storing 1,5 Mt/year CO2 in 2026 and up to 8 Mt/year CO2 in 2030. This corresponds to 13% of Denmark's annual CO2 emissions.
- In Belgium, the Antwerp port aims at capturing 50% of its emissions by 2030, which represents around 9 Mt/year CO2 to be stored in the North Sea.
- In France, the D'Artagnan project in Dunkerque is planned as an export hub opened to everyone to store CO2 in the North Sea from Dunkerque's harbour. The initial capacity is set at 3 Mt/year CO2 and will be increased up to 12 Mt/year CO2 in the final phase. Similar projects are proposed by Haropa Port for the Havre Port, to capture and store 7 Mt/year CO2.
- The UK still does not have any CCS projects but is aiming to develop them by 2030 to help reach climate neutrality. Indeed, the UK has considerable potential to store carbon under the North Sea. In the Net Zero Strategy, it is planned to capture and store between 20 Mt and 30 Mt/year CO2 from 2030; with capacity to be increased to 47 Mt/year CO2 from 2050.

Source: (Northern Lights, 2023[119]), (Porthos, 2023[120]), (Greensand, 2023[121]), (Port of Antwerp Bruges, 2023[122]), (Cornot-Gandolphe, 2021[123]), (Wettengel, 2023[124])

Demand for CCS is widely spread

Looking for potential CCS demand beyond Hamburg's immediate vicinity, European cement plants are geographically dispersed across the continent. Given that CCS infrastructure benefits from economies of scale, the wide distribution of cement facilities poses transportation challenges and raises costs. It may be that steel producers, with more concentrated production sites, are interested in joining cement production in CCS. Unit costs decline markedly as CO2 transport capacity increases.

Demand for CCS (Carbon capture and storage) in the Hamburg metropolitan area will come mainly from a cement plant nearby Hamburg in Lägerdorf. Copper production within Hamburg may add somewhat to local CCS demand (Table 3.5). Only part of these emissions would require CCS, as most are energy-related and more efficient production processes can reduce emissions.

The North Rhine-Westphalia region includes Europe's largest cement production clusters (Bellona Europa, 2016[125]), for a total CO2 emissions estimated at 35 Gt/year (IEA, 2020[60]). While CCS is still seen as a technology to be developed to limit the industrial emissions of the region beyond cement production

(Ministry for Economic Affairs, 2021[126]), uncertainties and the high cost of transporting CO2 over long distances make a role for Hamburg unlikely.

Figure 3.15. Scenarios for CO2 emissions of cement plants in Germany in 2050

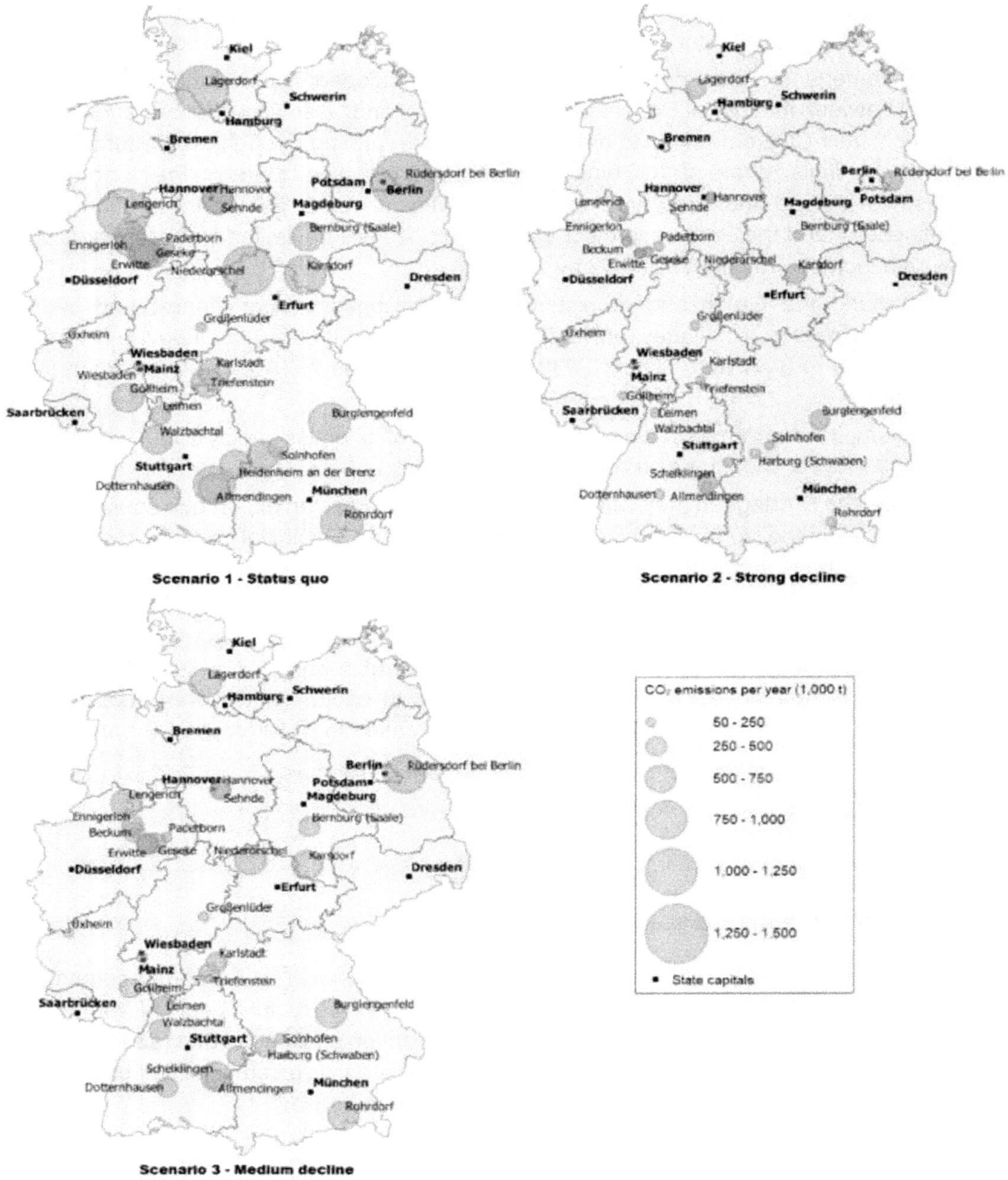

Note: Scenario 1 - No reduction in emissions until 2050, and cement demand and clinker-cement factor remain constant; Scenario 2 - 70% reduction in emissions until 2050, and use of alternative production processes and products; Scenario 3 - 35% reduction in emissions until 2050, and a smaller reduction in cement demand or CO2 emissions.
Source: (Winter, Schröter and Fidaschek, 2022[127])

Table 3.5. Cement and chemical CO2 emissions registered in the EU ETS for Hamburg

	Aurubis AG	Holcim (Deutschland) GmbH – Werk Lägerdorf (cement clinker)	Holcim (Deutschland) GmbH – Werk Höver (cement clinker)
Verified emissions (ETS) in 2022 (in tons)	154 294	961 548	555 043

Source: (EU ETS, 2022[128]), Aurubis AG.

The North Sea appears to be a preferred location for CCS

The IEA estimates theoretical storage capacity in the North Sea to be very large, around 300 Gt, equivalent to around 400 times German annual CO2 emissions, with almost half of them located in the Norwegian and British areas (IEA, 2022[129]) (Figure 3.16). Indeed, the North Sea is a mature oil and gas extraction area (Nakhle, 2016[130]), making available many pipelines and geological reservoirs that have been empty after decades of oil and gas development. These infrastructures could be restructured to carry CO2 emissions.

In addition to its substantial storage capacity, the oil and gas infrastructure in the North Sea presents key advantages over alternative storage locations. When considering the two primary options for CO2 storage saline aquifers and depleted oil and gas fields, it becomes evident that utilising oil and natural gas fields not only ensures long-term safeguards against the risk that CO2 re-emerges from storage sites (which would undo storage efforts) but also provides significantly more precise capacity estimates, allowing to assess the economic viability of each storage site. The North Sea region thus enhances the overall viability and reliability of carbon capture and storage efforts (OECD, 2023[43]).

One key factor favouring offshore storage is social acceptance. The German government will soon unveil its carbon capture and storage strategy (BMWK, 2022[131]). There are plans to introduce a reform project in 2023 that will enable the construction of commercial infrastructure for CCS. While onshore storage remains prohibited in all *Länder*, offshore storage in the North Sea is permitted. For now, in Germany, no storage or pipelines have been applied for, approved or built (Wettengel, 2023[124]).

Figure 3.16. Potential offshore CCS storage sites in North Sea hydrocarbon fields vary in storage capacity

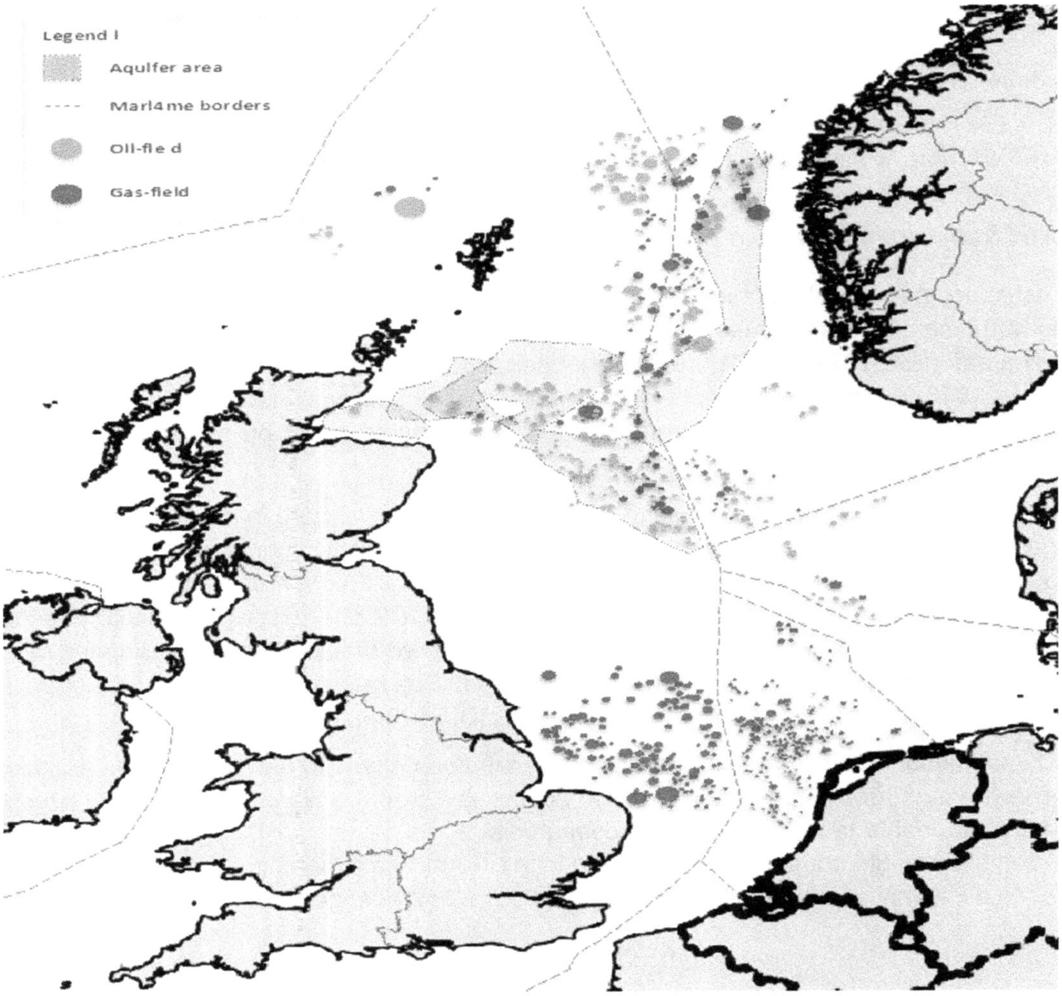

Source: NLOG. (2020). Interactieve Kaart. https://www.nlog.nl/ NPD (2020). Interactive map — Norwegianpetroleum.https://www.norskpetroleum.no/en/interactive-map-quick-downloads/interactive-map/ . OGA. (2020). Interactive maps and tools. https://www.ogauthority.co.uk/data-centre/interactive-maps-and-tools/

High costs and uncertainty make a major role of Hamburg in CCS unlikely

CCS is an exceptionally expensive procedure. At present, based on data from the interactive tool created by the Clean Air Task Force, the cost of CCS in Germany varies significantly, ranging from EUR 70 to EUR 280 per ton of CO2, with an average cost of EUR 211 per ton for the cement industry (Clean Air Task Force, 2023[132]). In contrast, the current market price for a ton of CO2 hovers at approximately EUR 80, making CCS economically unfeasible.

Research and development efforts have the potential to significantly decrease costs associated with large-scale CCS implementation (Budinis et al., 2018[133]). The areas of CCS research are vast, ranging from reducing energy requirements and using innovative solvents to standardising capture units and increasing plant size. They could reduce costs by 80% (IEA, 2020[60]). Furthermore, costs are expected to fall as the market expands (Baylin-Stern and Berghout, 2021[134]). Overall, the cost of CCS should decrease, but it is heavily contingent on research and development efforts, which introduce a significant level of uncertainty.

Transportation of CO2 emissions is complex and may not require port infrastructure

Emission sources that could potentially utilise CCS are often located at a considerable distance from the North Sea, particularly in the case of widely dispersed cement production facilities, raising two main transportation challenges: onshore transport for CO2 collection and offshore transport to deliver the CO2 to storage sites.

First, with respect to onshore transportation, the most likely economically viable transportation is an onshore pipeline. Only when the quantity or distance of CO2 transported is limited, CO2 emissions can be transported over land via train or truck. Presently such a pipeline network does not exist; if a pipeline network is built to connect emission sites with the North Sea, it may not go through Hamburg. Contrary to hydrogen, the transport of CO2 cannot easily use repurposed natural gas pipelines. It requires new infrastructure which is time-consuming and costly (OECD, 2023[43]). Building an onshore pipeline for CO2 emissions transport can present big challenges and high costs, particularly due to the potential for land use conflicts.

For beneath the North Sea, in terms of cost-effectiveness, offshore pipelines are well-suited for large quantities and relatively short distances, whereas shipping emerges as a more financially viable choice for conveying smaller quantities of CO2 across extended distances (Figure 3.17) (IEA, 2020[135]).

Figure 3.17. Shipping and offshore pipeline transportation costs

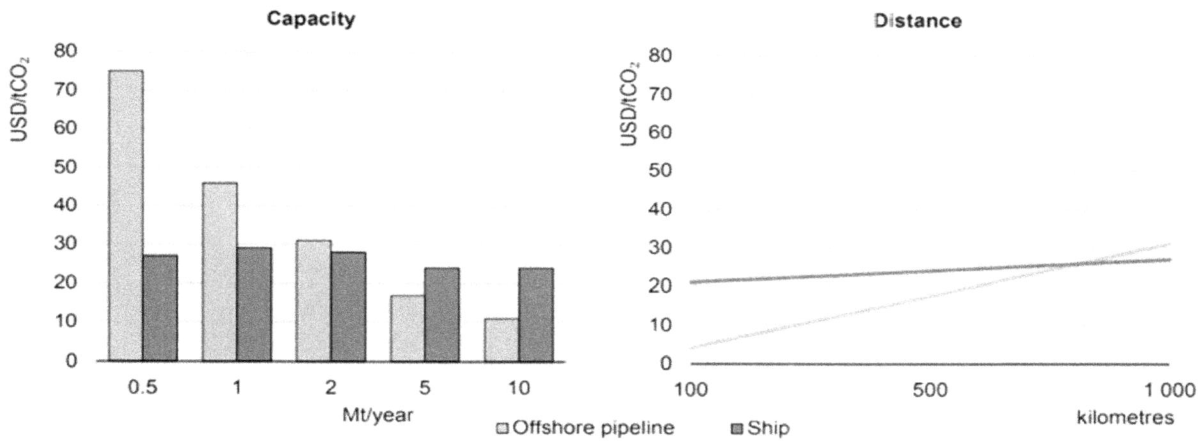

Note: Mt denotes million tons. The left-hand chart assumes a distance of 1 000 km. The right-hand chart assumes a capacity of 2 Mt/year.
Source: (IEA, 2020[61])

Several initiatives are emerging to connect with North Sea storage facilities via offshore pipelines. They enable the industrial-scale transportation of CO2, are less vulnerable to disruptions caused by extreme weather conditions, and do not necessitate any additional technological advancements. Offshore pipelines may not require a dedicated port facility such as Hamburg's (OECD, 2023[43]).

Several CCS projects in Europe are planning to use shipping as the primary form of transport, which is more flexible (IEA, 2022[129]). Equinor plans to build two ships that can transport liquefied gas: it will collect CO2 industrial emissions from Northern Europe areas close to ports like Hamburg and will go afterwards to Oygarden, an offshore storage site on the Norwegian coast (Joeres, 2021[136]).

Instead of being stored, CO2 can be used locally in a number of industries, notably in basic chemical manufacturing, such as the production of plastics, lubricants or waxes, which are produced also in Hamburg. In combination with hydrogen, CO2 could replace fossil fuel feedstocks in these production processes. It could for example be employed in the production of green methanol, which is likely to play a role as a fuel in maritime shipping. To limit cost, the plant using the CO2 would need to be close to the CO2 capture site while supply and demand of CO2 would need to match closely to remain consistent with climate neutrality. Moreover, to be fully consistent with climate neutrality, captured CO2 emissions would need to be used in products that do not themselves emit CO2 in use or disposal. Otherwise, the CO2 would need to come from non-emitting CO2 sources, such as the firing of sustainable biomass or direct air capture (DAC). CCU and DAC are energy intensive, which could undermine the energy transition (OECD, 2023[43]). Overall, the potential for CCU may be limited.

Box 3.13. Current offshore CO2 pipeline systems in Europe connecting the mainland to North Sea storage sites

In Norway, the Hammerfest is 153 km long and has a capacity of 0,7 Mt/year. In the Netherlands, the Rotterdam system is 85 km long with a capacity of 0,4 Mt/year (IEA, 2020[60]). Germany's proximity with south Norwegian EOR fields is an advantage to store CO2 in the North Sea (Global CCS institute, 2007[137]). A third CO2 pipeline may be built between Germany and Norway, to store CO2 in the North Sea, and may not be built from Hamburg (Figure 3.18). This subsea pipeline would be 900 km long, with a capacity between 20 and 40 Mt/year by 2037, equivalent to around 20% of Germany's annual industrial emissions (Offshore, 2022[138]). It may compete with other locations, such as Rotterdam.

Figure 3.18. Pipeline project to transport CO2 between Germany and Norway

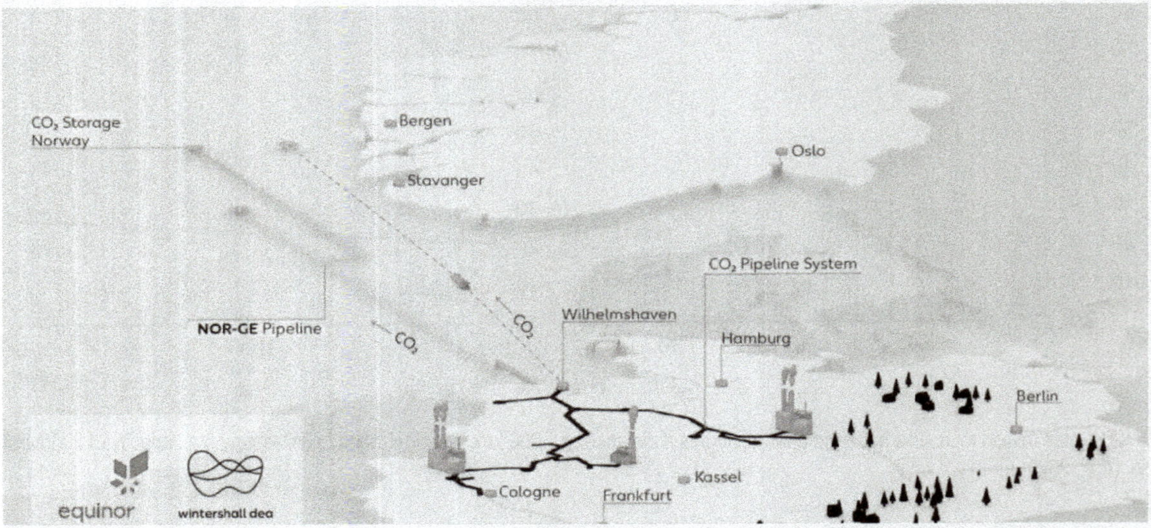

Notes: EOR stands for Enhanced oil recovery, a process for extracting oil. Wintershall Dea is a German gas and oil company and Equinor is a Norwegian gas and oil company.
Source: (Wintershall Dea, 2022[139])

The port may support local industries with CCS needs

To conclude, Hamburg's potential to establish itself as a prominent hub for CO2 collection appears, at best, uncertain and, at worst, improbable. Rotterdam would be a closer location as a hub for storing CO2 from North-Western Europe below the North Sea. Hamburg's port infrastructure is more likely to offer CCS-related services to the relatively modest emissions in the Hamburg region which may require CCS, notably for cement production nearby Hamburg and copper production in Hamburg. Hamburg's direct access to the North Sea presents opportunities for directly sending locally produced CO2 to storage sites.

Key actions

By 2030

- Businesses should identify needed CCS-related services, notably from local cement and copper production

By 2040

- Businesses should develop any needed port CCS-related services to local industry and local businesses bearing in mind the availability of space or transhipment facilities, as well as the compatibility with other transformation developments.

References

ACEA (2022), *European EV Charging Infrastructure Masterplan*. [24]

Acerinox (2021), *Annual Integrated Report 2020*, https://www.acerinox.com/opencms901/export/sites/acerinox/.content/galerias/galeria-descargas/InformeIntegradoACX2020last.pdf (accessed on 30 July 2021). [62]

ArcelorMittal (2021), *Climate Action Report 2*, https://corporate-media.arcelormittal.com/media/ob3lpdom/car2.pdf (accessed on 30 July 2021). [63]

ArcelorMittal (2021), *German Federal Government commits its intention to provide €55 million of funding for ArcelorMittal's Hydrogen DRI plant*, https://corporate.arcelormittal.com/media/news-articles/german-federal-government-commits-its-intention-to-provide-55-million-of-funding-for-arcelormittal-s-hydrogen-dri-plant (accessed on 27 September 2021). [69]

Aurubis AG (2023), *2023 Sustainability Report*. [81]

BALM (2023), *Marktbeobachtung Güterverkehr*. [12]

Barteková, E. and P. Börkey (2022), "Digitalisation for the transition to a resource efficient and circular economy", *OECD Environment Working Papers*, No. 192, OECD Publishing, Paris, https://doi.org/10.1787/6f6d18e7-en. [53]

Bataille, C. (2020), *Low and zero emissions in the steel and cement industries: Barriers, technologies and policies", OECD Green Growth Papers, No. 2020/02*, OECD Publishing, Paris, https://doi.org/10.1787/5ccf8e33-en. [57]

Baylin-Stern, A. and N. Berghout (2021), "Is carbon capture too expensive?". [134]

Behörde für Wirtschaft und Innovation (2023), *Hafenentwicklungsplan*, https://www.hamburg.de/bwi/medien/17188570/2023-06-13-bwi-hafenentwicklungsplan/ (accessed on 2023). [2]

Bellona Europa (2016), *Manufacturing Our Future: Industries, European Regions, and Climate Action*. [125]

BMDV (2022), *Charging Infrastructure Masterplan II*. [26]

BMWK (2022), *Evaluierungsbericht der Bundesregierung zum Kohlendioxid-Speicherungsgesetz (KSpG)*. [131]

Budinis, S. et al. (2018), "An assessment of CCS costs, barriers and potential", *Energy Strategy Reviews*, Vol. 22, pp. 61-81, https://doi.org/10.1016/J.ESR.2018.08.003. [133]

BWI (2023), *2040 Port Development Plan*. [16]

Camisón-Haba and Clemente-Almendros (2020), *Hydrogen – the drive of the future?*. [140]

Carbon Disclosure Project (2022), *Aurubis AG - Climate Change 2022*. [82]

Cedefop (2021), *Digital, greener and more resilient. Insights from Cedefop's European skills forecast*, Luxembourg:, https://www.cedefop.europa.eu/files/4201_en.pdf;http://data.europa.eu/doi/10.2801/154094. [87]

CertifHy (2023), *CertifHy™ has submitted its RFNBO Voluntary Scheme for recognition by the European Commission*, http://www.certifhy.eu. [116]

Chen, J. et al. (2019), "Environmental benefits of secondary copper from primary copper based on life cycle assessment in China", *Resources, Conservation and Recycling*, Vol. 146, pp. 35-44, https://doi.org/10.1016/j.resconrec.2019.03.020. [79]

Clean Air Task Force (2023), , https://www.catf.us/ccs-cost-tool/. [132]

Collins, L. (2023), *German cabinet approves creation of a national core hydrogen pipeline network, to become law this year*. [113]

Cornot-Gandolphe, S. (2021), "Un nouvel élan pour le captage, stockage et utilisation du carbone (CCUS) en Europe". [123]

Dachser (2022), *Hydrogen – the Drive of the Future?*. [29]

de Leeuw, P. and S. Kim (2021), *UK offshore energy workforce transferability review*, Robert Gordon University, https://www.rgueti.com/wp-content/uploads/2021/05/workforce-transferability-report.pdf. [93]

Elshkaki, A. et al. (2016), "Copper demand, supply, and associated energy use to 2050", *Global Environmental Change*, Vol. 39, pp. 305-315, https://doi.org/10.1016/j.gloenvcha.2016.06.006. [71]

ENTSO-E (2014), *D 2.2 European cluster model of the Pan-European transmission grid. e-HIGHWAY 2050 - Modular Development Plan of the Pan-European Transmission System 2050.*, https://docs.entsoe.eu/baltic-conf/bites/www.e-highway2050.eu/fileadmin/documents/Results/D2_2_European_cluster_model_of_the_Pan-European_transmission_grid_20072015.pdf. [106]

Equitable Maritime Consulting (2023), *Evidence based policy advisory service*. [7]

ERA (2022), *Report Fostering the Railway Sector Through the European Green Deal: Rail-Port Synergies*, European Union Agency for Railways, Valenciennes, https://www.era.europa.eu/system/files?file=2023-01/fostering_the_railway_sector_through_the_european_green_deal_-_rail-ports_synergies_1.pdf. [1]

Erbach, G. and S. Svensson (2023), *BRIEFING Towards climate neutrality*. [115]

EU ETS (2022), *European Union Transaction Log*. [128]

Eurofer (2022), *European Steel in Figures - Employment and Economic Impact*. [58]

European Aluminium (2015), *The European aluminium industry's sustainability roadmap towards 2025*, https://european-aluminium.eu/media/1034/sustainability-roadmap.pdf. [46]

European Commission (2022), *EU Transport in Figures – Statistical pocketbook 2022, Directorate-General for Mobility and Transport*. [31]

European Commission (2022), *Key actions of the EU Hydrogen Strategy*. [97]

European Commission (2022), *REPowerEU - Une énergie abordable, sûre et durable pour l'Europe*. [98]

Eurostat (2023), *Eurostat Data Browser*. [34]

Federal Ministry for Economic Affairs and Energy (2020), *The National Hydrogen Strategy*. [99]

Fennell, P. et al. (2022), "Cement and steel — nine steps to net zero", *Nature*, Vol. 603/7902, pp. 574-577, https://doi.org/10.1038/d41586-022-00758-4. [70]

Fraunhofer Institute (2022), *Endbericht Machbarkeitsstudie Water Cargo Barge (WACABA)*. [21]

Fraunhofer Institute et al. (2022), *Mögliche Ausbauschritte für eine Oberleitungsinfrastruktur für den Straßengüterverkehr in Deutschland, Eine kriterienbasierte Analyse*. [27]

Future Hamburg (2022), *"Growing demand: H2 - Future in the making"*. [101]

Future Hamburg (2022), *Pilot plant in Hamburg: Steel goes green*. [104]

Global CCS institute (2007), *Development of a CO2 transport and storage network in the North Sea*. [137]

Greensand (2023), *Project Greensand | CO2 Lagring*, https://www.projectgreensand.com/en (accessed on 19 July 2023). [121]

Grimes, S., J. Donaldson and J. Grimes (2015), *Report on the Environmental Benefits of Recycling - 2016 edition*, Recycling, Bureau of International, https://www.bir.org/publications/facts-figures/download/172/174/36?method=view. [73]

H2 Global Stiftung (2023), *The H2Global Instrument*. [108]

Halim, R. et al. (2018), "Decarbonization Pathways for International Maritime Transport: A Model-Based Policy Impact Assessment". [3]

Halim, R., T. Smith and D. Englert (2019), "Understanding the Economic Impacts of Greenhouse Gas Mitigation Policies on Shipping - What Is the State of the Art of Current Modeling Approaches?". [9]

Hamburg Invest (2022), *A strong hydrogen region*. [102]

Hamburgnews (2020), "HY-5 on way to becoming Europe's leading hydrogen region". [103]

HHLA (2023), *HHLA Pure*. [17]

HHLA (2022), *Hydrogen as a future opportunity*. [105]

Hoang, A. et al. (2022), "Energy-related approach for reduction of CO2 emissions: A critical strategy on the port-to-ship pathway", *Journal of Cleaner Production*, Vol. 355, p. 131772, https://doi.org/10.1016/j.jclepro.2022.131772. [6]

Hund, K. et al. (2023), *Minerals for Climate Action: The Mineral Intensity of the Clean Energy Transition*, World Bank, Washington, DC, https://openknowledge.worldbank.org/handle/10986/40002. [75]

Hydrogeninsight (2023), "Green hydrogen imported to Europe would be cost-competitive with locally produced H2 by 2030: analyst". [112]

IAI (2021), *Aluminium Sector Greenhouse Gas Pathways to 2050*, The International Aluminium Institute, London, https://www.world-aluminium.org/media/filer_public/2021/03/16/iai_ghg_pathways_position_paper.pdf. [86]

ICCT (2023), *European Union Alternative Fuel Infrastructure Regulation (AFIR)*, The International Council for Clean Transportation, Berlin. [25]

ICCT and ECTA (2022), *Readiness of The European Fleets for Zero-Emission Trucking*. [42]

IEA (2023), *Hydrogen*. [95]

IEA (2022), *Global Hydrogen Review 2022*. [96]

IEA (2022), *Regional opportunities*. [129]

IEA (2021), *How Energy Efficiency Will Power Net Zero Climate Goals*. [41]

IEA (2021), *Net Zero by 2050: A Roadmap for the Global Energy Sector*, OECD Publishing, Paris, https://doi.org/10.1787/c8328405-en. [90]

IEA (2021), *The Role of Critical Minerals in Clean Energy Transitions*, OECD Publishing, Paris, https://doi.org/10.1787/f262b91c-en. [80]

IEA (2020), *CCUS technology innovation – CCUS in Clean Energy Transitions – Analysis - IEA*, https://www.iea.org/reports/ccus-in-clean-energy-transitions/ccus-technology-innovation (accessed on 18 July 2023). [135]

IEA (2020), *Energy Technology Perspectives 2020*, OECD Publishing, Paris, https://doi.org/10.1787/d07136f0-en. [61]

IEA (2020), *Energy Technology Perspectives 2020 - Special Report on Carbon Capture Utilisation and Storage : CCUS in clean energy transitions*, OECD Publishing, Paris, https://doi.org/10.1787/208b66f4-en. [60]

IEA (2020), *Iron and Steel Technology Roadmap: Towards more sustainable steelmaking*, OECD Publishing, Paris, https://doi.org/10.1787/3dcc2a1b-en. [55]

IEA (2019), *"The Future of Hydrogen - Seizing today's opportunities"*. [100]

IMO (2023), *2023 IMO Strategy on Reduction of GHG Emissions from Ships*. [4]

International Copper Association (2023), *Copper - The Pathway to Net Zero*, https://copperalliance.org/wp-content/uploads/2023/03/ICA-GlobalDecar-202301-English-Final-singlepgs.pdf. [72]

International PtX Hub (2023), *Catalysing defossilisation globally*. [107]

Intraplan Trimode (2023), *Gleitende Langfrist-Verkehrsprognose 2021-2022*. [14]

IPCC (2022), *Industry*. [118]

IRENA (2022), *"Geopolitics of the Energy Transformation - The Hydrogen Factor"*. [111]

IRENA (2022), *Global hydrogen trade to meet the 1,5°C climate goal*. [110]

IRENA (2020), *"Green hydrogen - A guide to policy making"*. [94]

IRENA coalition for action (2021), "DECARBONISING END-USE SECTORS: PRACTICAL INSIGHTS ON GREEN HYDROGEN". [117]

IRENA and RMI (2023), *CREATING A GLOBAL HYDROGEN - MARKET CERTIFICATION TO ENABLE TRADE*. [114]

ITF (2023), *Decarbonisation and the Pricing of Road Transport: Summary and Conclusions*, OECD Publishing, Paris. [30]

ITF (2023), *ITF Transport Outlook 2023*. [35]

ITF (2022), *Decarbonising Europe's Trucks - How to Minimise Cost Uncertainty*. [18]

ITF (2022), *How Digitally-driven Operational Improvements Can Reduce Global Freight Emissions*. [20]

ITF (2022), *Mode Choice in Freight Transport*. [23]

ITF (2021), *Decarbonising Transport in Europe - The Way Forward*. [13]

ITF (2021), *Zero Carbon Supply Chains: The Case of Hamburg*, OECD Publishing, Paris. [38]

ITF (2020), *How Urban Delivery Vehicles Can Boost Electric Mobility*, OECD Publishing, Paris. [33]

Jing, L. et al. (2020), "Carbon intensity of global crude oil refining and mitigation potential", *Nature Climate Change*, Vol. 10, pp. 526–532, https://doi.org/10.1038/s41558-020-0775-3. [89]

Joeres, A. (2021), "Les fausses promesses des technologies de captage du carbone pour réduire les émissions de CO2", *Le Monde*. [136]

Johnson, O. et al. (2021), "Toward Climate-Neutral Heavy Industry: An Analysis of Industry Transition Roadmaps", *Applied Sciences*, Vol. 11/12, p. 5375, https://doi.org/10.3390/app11125375. [48]

Karka, P., F. Johnsson and S. Papadokonstantakis (2021), “Perspectives for Greening European Fossil-Fuel Infrastructures Through Use of Biomass: The Case of Liquid Biofuels Based on Lignocellulosic Resources”, *Frontiers in Energy Research*, Vol. 9, https://doi.org/10.3389/fenrg.2021.636782. [91]

LIHH (2023), *Logistik Report 2022-2023 für die Metropolregion Hamburg*. [32]

Liu, G., C. Bangs and D. Müller (2013), “Stock dynamics and emission pathways of the global aluminium cycle”, *Nature Climate Change*, Vol. 3, pp. 338–342, https://doi.org/10.1038/nclimate1698. [85]

Liu, G. and D. Müller (2012), “Addressing sustainability in the aluminum industry: a critical review of life cycle assessments”, *Journal of Cleaner Production*, Vol. 35, pp. 108-117, https://doi.org/10.1016/j.jclepro.2012.05.030. [84]

Luidold, S. and H. Antrekowitsch (2007), “Hydrogen as a reducing agent: State-of-the-art science and technology”, *JOM*, Vol. 59/6, pp. 20-26, https://doi.org/10.1007/s11837-007-0072-x. [78]

Mærsk Mc-Kinney Møller Center (2021), *Decarbonizing the global maritime industry*. [5]

Material Economics (2019), *Industrial Transformation 2050 - Pathways to Net-Zero Emissions from EU Heavy Industry*, University of Cambridge Institute for Sustainability Leadership (CISL), https://materialeconomics.com/publications/industrial-transformation-2050. [44]

Material Economics (2018), *Retaining value in the Swedish Materials System*, https://circulareconomy.europa.eu/platform/en/main-language/swedish. [51]

McKinsey & Company (2022), *Bold Moves to Boost European Freight*. [37]

Merk and Notteboom (2015), *Port Hinterland Connectivity*, OECD Publishing, Paris. [19]

Ministry for Economic Affairs, I. (2021), *“Carbon management for climate protection - The Carbon Management Strategy of North Rhine-Westphalia”*. [126]

Nakhle, C. (2016), *News Assessing the Future of North Sea Oil and Gas*, Carnegie Middle East Center. [130]

Northern Lights (2023), *What we do - Northern Lights*, https://norlights.com/what-we-do/ (accessed on 19 July 2023). [119]

OECD (2023), *Regional Industrial Transitions to Climate Neutrality*, OECD Regional Development Studies, OECD Publishing, Paris, https://doi.org/10.1787/35247cc7-en. [43]

OECD (2020), *The Circular Economy in Cities and Regions: Synthesis Report*, OECD Urban Studies, OECD Publishing, Paris, https://doi.org/10.1787/10ac6ae4-en. [52]

OECD (2019), *Business Models for the Circular Economy: Opportunities and Challenges for Policy*, OECD Publishing, Paris, https://doi.org/10.1787/g2g9dd62-en. [54]

OECD (2019), *Global Material Resources Outlook to 2060: Economic Drivers and Environmental Consequences*, OECD Publishing, Paris, https://doi.org/10.1787/9789264307452-en. [50]

OECD/Hywax (2023), *OECD interview with Hywax*. [92]

Offshore (2022), *North Sea pipeline to transport CO2 from Germany to storage sites in Norway*. [138]

Oxford Economics (2019), *The impact of the European steel industry on the EU economy*, https://www.eurofer.eu/assets/Uploads/20190530-OE-report-for-Eurofer-1-v2.pdf. [45]

Pedneault, J. et al. (2021), "What future for primary aluminium production in a decarbonizing economy?", *Global Environmental Change*, Vol. 69, p. 102316, https://doi.org/10.1016/j.gloenvcha.2021.102316. [83]

Pei, M. et al. (2020), "Toward a Fossil Free Future with HYBRIT: Development of Iron and Steelmaking Technology in Sweden and Finland", *Metals*, Vol. 10/7, p. 972, https://doi.org/10.3390/met10070972. [67]

Port of Antwerp Bruges (2023), *Antwerp@C investigates potential for halving CO2 emissions in Port of Antwerp by 2030*. [122]

Port of Barcelona (2019), *eCOcalculator*. [40]

Port of Halifax (2023), *Annual Report for 2022*. [39]

Port of Hamburg Marketing (2023), "Charts, Yearly Press Conference 2023". [10]

Porthos (2023), *CO2 reduction through storage under the North Sea - Porthos*, https://www.porthosco2.nl/en/ (accessed on 19 July 2023). [120]

Posco (2021), *Posco Corporate Citizenship Report 2020*, https://www.posco.co.kr/docs/eng6/jsp/dn/irinfo/posco_report_2020.pdf#page=1 (accessed on 30 July 2021). [64]

Quader, A. et al. (2016), "Present needs, recent progress and future trends of energy-efficient Ultra-Low Carbon Dioxide (CO2) Steelmaking (ULCOS) program", *Renewable and Sustainable Energy Reviews*, Vol. 55, pp. 537-549, https://doi.org/10.1016/j.rser.2015.10.101. [66]

REH (2023), *Hydrogen Economy in the Hamburg Metropolitan Region*. [28]

Reuters (2021), *ArcelorMittal gets support for green steel plant in Hamburg*, https://www.reuters.com/business/sustainable-business/arcelormittal-gets-support-green-steel-plant-hamburg-2021-09-07/ (accessed on 27 September 2021). [59]

Sharmina, M. et al. (2021), "Decarbonising the critical sectors of aviation, shipping, road freight and industry to limit warming to 1.5–2°C", *Climate Policy*, Vol. 21/4, pp. 455-474, https://doi.org/10.1080/14693062.2020.1831430. [49]

SSAB (2021), *The world's first fossil-free steel ready for delivery*, https://www.ssab.com/News/2021/08/The-worlds-first-fossilfree-steel-ready-for-delivery (accessed on 27 September 2021). [68]

Sun, X., F. Lettow and K. Neuhoff (2021), "Climate neutrality requires coordinated measures for high quality recycling", *DIW Weekly Report*, Vol. 6, https://doi.org/10.18723/diw_dwr:2021-26-1. [47]

T&E (2021), "Unlocking Electric Trucking in the EU: Long-Haul Trucks". [11]

The Royal Society (2021), *The role of hydrogen and ammonia in meeting the net zero challenge*, https://royalsociety.org/-/media/policy/projects/climate-change-science-solutions/climate-science-solutions-overview.pdf. [77]

Thyssenkrupp (2020), *Annual Report 2019/2020*, https://ucpcdn.thyssenkrupp.com/_binary/UCPthyssenkruppAG/en/investors/link-thyssenkrupp-GB-2019-2020-ENG-Web.pdf (accessed on 30 July 2021). [65]

Toll Collect (2023), *Toll Rates per km from1 January 2023.* [36]

UBA (2023), *UBA forecast: 2022 Greenhouse Gas Emissions Down by 1.9 percent.* [15]

University of Cambridge Institute for Sustainability Leadership (CISL) (2020), *Working towards a climate neutral Europe: Jobs and skills in a changing world*, Technical Report. Cambridge, UK: CLG Europe, https://www.corporateleadersgroup.com/reports-evidence-and-insights/technical-report-working-towards-a-climate-neutral-europe-jobs-and-skills-in-a-changing-world (accessed on 12 October 2021). [88]

Valera-Medina, A. et al. (2018), “Ammonia for power”, *Progress in Energy and Combustion Science*, Vol. 69, pp. 63-102, https://doi.org/10.1016/j.pecs.2018.07.001. [76]

VDA (2022), *Decarbonisation of road freight transport Suitable framework conditions and rapid implementation of the Charging Infrastructure Master Plan II.* [22]

Wang, P. et al. (2021), “Efficiency stagnation in global steel production urges joint supply- and demand-side mitigation efforts”, *Nature Communications*, Vol. 12/1, pp. 1-11, https://doi.org/10.1038/s41467-021-22245-6. [56]

Watari, T. et al. (2022), “Global copper cycles and greenhouse gas emissions in a 1.5 °C world”, *Resources, Conservation and Recycling*, Vol. 179, p. 106118, https://doi.org/10.1016/j.resconrec.2021.106118. [74]

Weltenergierat (2020), *International hydrogen strategies.* [109]

Wettengel, J. (2023), *Quest for climate neutrality puts CCS back on the table in Germany*, Clean Energy Wire. [124]

Winter, C., B. Schröter and S. Fidaschek (2022), *The German Cement Industry as a CO2 Source for Other Industries.* [127]

Wintershall Dea (2022), *Wintershall Dea and Equinor partner up for large-scale CCS value chain in the North Sea.* [139]

World Bank (2021), *Charting a Course for Decarbonizing Maritime Transport*, https://www.worldbank.org/en/news/feature/2021/04/15/charting-a-course-for-decarbonizing-maritime-transport. [8]

Annex 3.A. Description of models and further modelling results

The NavigaTe, IFM, and GEM-E3 models

NavigaTE model: Techno-economic model to assess potential decarbonisation pathways for shipping, which captures the entire maritime energy value chain for powering the vessel from feedstock/primary energy to the wake of the vessel. The NavigaTE model consists of two main elements: A Total Cost of Ownership (TCO) model and an Industry Transition model. These are used to estimate the fleet fuel uptake over time and energy efficiencies leading to an overall estimate of the required energy demand, fuel split and GHG emissions.

International freight transport demand model (IFM): a freight transport model specifically designed to project international freight transport activities (in tonne-kilometres), maritime transport costs and modal share until 2050 under different trade and policy scenarios. EMC hosts one of the leading international freight transport models that has been validated, published in a scientific journal1 and applied to project transport activities in different maritime decarbonisation projects, including IMO impact assessments. The model is built on the four-step freight transportation modelling approach and it takes the global trade projection from the GEM-E3 model as an input. The IFM is designed to be able to estimate the weight of commodities traded between countries, the choice between modes and transport routes used to transport these commodities based on transport network characteristics, and relevant socio-economic variables such as transport costs and time. The model consists of the following components:

1. Trade flow disaggregation model;
2. Value-to-weight model;
3. Mode choice model; and
4. Route choice model.

The model is highly granular with spatial resolution that covers more than 120 countries, 333 regions, and 4000 ports worldwide. Trade flow disaggregation and value-to-weight sub-models allow the model to convert international trade values (typically presented in USD in various global trade databases) into trade volume (in tonnes) and assign them across global maritime routes using the route choice model and assignment procedure. The mode and route choice sub-models are also capable of analysing the potential shift to other available modes and routes as an impact of changes in transport costs.

The output generated by IFM is the bilateral trade value of each commodity at the centroid level (i.e., city-level) by each transportation mode. This means that in order to evaluate the impact on an individual city (i.e., Hamburg) in terms of its maritime imports/exports per commodity and its maritime trade relationship with important trading partners at the country level, there's a need to gather and summarize information at both the product and country levels. As a result, both disaggregation and aggregation processes are conducted.

In the step of computing Hamburg's imports/exports maritime unit transport costs per commodity from/to all trading partner countries, we apply a weighted average approach to aggregate information.

In this procedure, the maritime unit transport cost of each import/export to/from Hamburg is assigned a weight corresponding to the ratio of that import/export flow to the total import/export volume of the port, distinguished by commodity and year. The weighted average value is then obtained by summing the weighted unit transport costs of all imports/exports. The Formula (1) is given to measure the weighted average export unit transport costs for Hamburg. The approach to calculating the weighted average for import unit transport costs is similar, by only considering all the origins to Hamburg.

$$W_{dcy} = \frac{V_{dcy}}{\sum_{d=1}^{D} V_{dcy}}$$

$$WUTC_{cy} = \sum_{d \in D} (W_{dcy} * UTC_{dcy}) \qquad (1)$$

In Equation (1):

V_{dcy} = Trade value from Hamburg to destination centroid *d* for commodity *c* in year *y*,

W_{dcy} = Weight of trade from Hamburg to destination centroid *d* for commodity *c* in year *y*,

UTC_{dcy} = Unit transport cost between Hamburg and destination centroid *d* (US$/ton-km),

$WUTC_{cy}$ = Weighted export unit transport costs for Hamburg for commodity *c* in year *y*.

GEM-E3: A global economic model designed to simulate the operation of the economic system (by country) with a particular focus on the representation of bilateral trade transactions by origin-destination, product and transport model. Focus is placed on the global trade volume of commodities transported using maritime transport. GEM-E3 is a state-of-the-art model (accumulating knowledge of over 20 years of model development – it is continuously used to support impact assessment of a range of policies) which is capable of simulating the interdependencies of markets and economic agents' decisions at a global scale. The model captures changes in global trade that are driven by competitiveness, policies/regulations/standards, infrastructure, prices, and supply/demand constraints. The representation of commodity trade includes numerous factors that are relevant for policy analysis such as maritime and international transport costs, market-based instruments such as carbon tax, and the characteristics of different countries in producing supply and demand for commodities globally.

A 3-step modelling approach

First, NavigaTe estimates the increase in total capital expenditures (CAPEX) and operational expenditures, (OPEX) taking into account the Well-to-Wake life cycle costs of production and deployment of zero-carbon fuels such as green methanol, and ammonia for different types of vessels due to the implementation of the GFS and the carbon levy in 2030, 2040, and 2050.

Second, based on the estimated increase in vessel costs, IFM translates this impact to the increase in maritime unit transport costs across all commodities and shipping routes globally. The computation takes into account the volume of freight transported across different routes and different cost components across the intermodal transport chain including hinterland and sea transport, to estimate the increase in maritime transport costs.

Third, GEM-E3 takes the increase in maritime transport costs to estimate the shift in trade patterns between countries, as reflected in the change in import/export values of all commodities by maritime mode into/out of Hamburg. Changes in values include both changes in volumes as well as changes in relative prices. The impact of the EC proposals for the IMO decarbonisation measures can be inferred from the changes in sea trade value by comparing the change in the import/export values between BAU and policy scenarios for each commodity.

Because GEM-E3's results are presented at a broader regional scale, combining multiple countries within a single region, it is necessary to break down these regional trade patterns into trading relationships of countries and port cities. To achieve this, the trade disaggregation module of IFM is utilised.

To assess the potential second-order feedback effect of change in trade value on maritime unit transport costs, for example, due to economies of scale, the analysis deploys this change in trade volume to adjust the transport costs estimated by IFM. Specifically, a trade transport cost elasticity as specified in the study of (Camisón-Haba and Clemente-Almendros, 2020[140]) is applied.

Further modelling results

Annex Figure 3.A.1. Projected increase in unit transport costs of each commodity imported to Hamburg (%)

Percent change (%) in import unit transport costs (USD/ton-km), relative to business-as-usual scenario

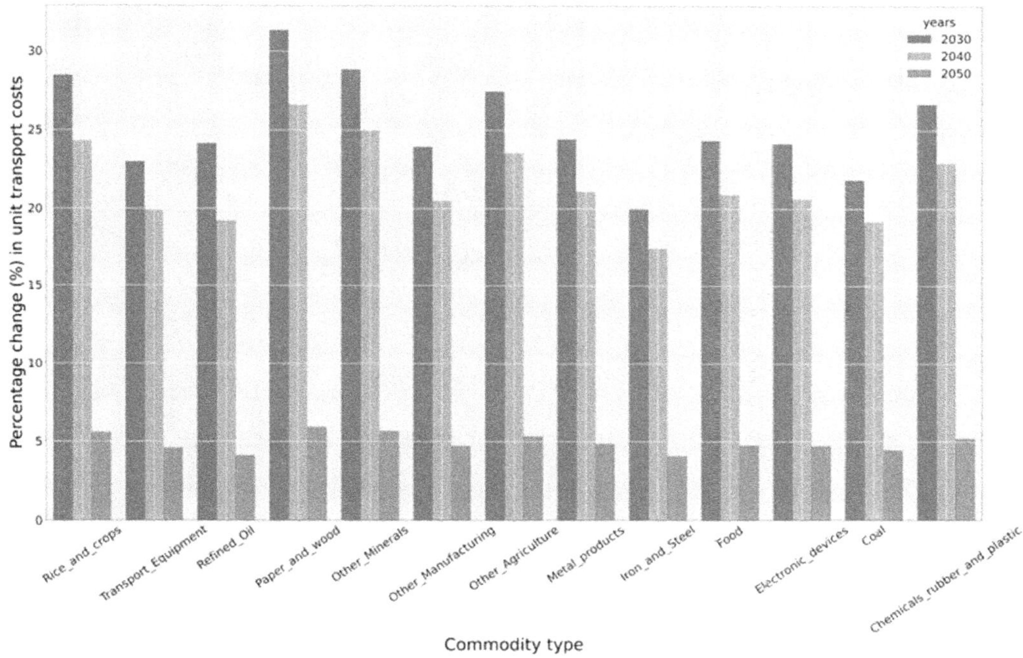

Source: (Equitable Maritime Consulting, 2023[7])

Annex Figure 3.A.2. Projected increase in unit transport costs of each commodity exported from Hamburg (%)

Percent change (%) in export unit transport costs (USD/ton-km), relative to business-as-usual scenario

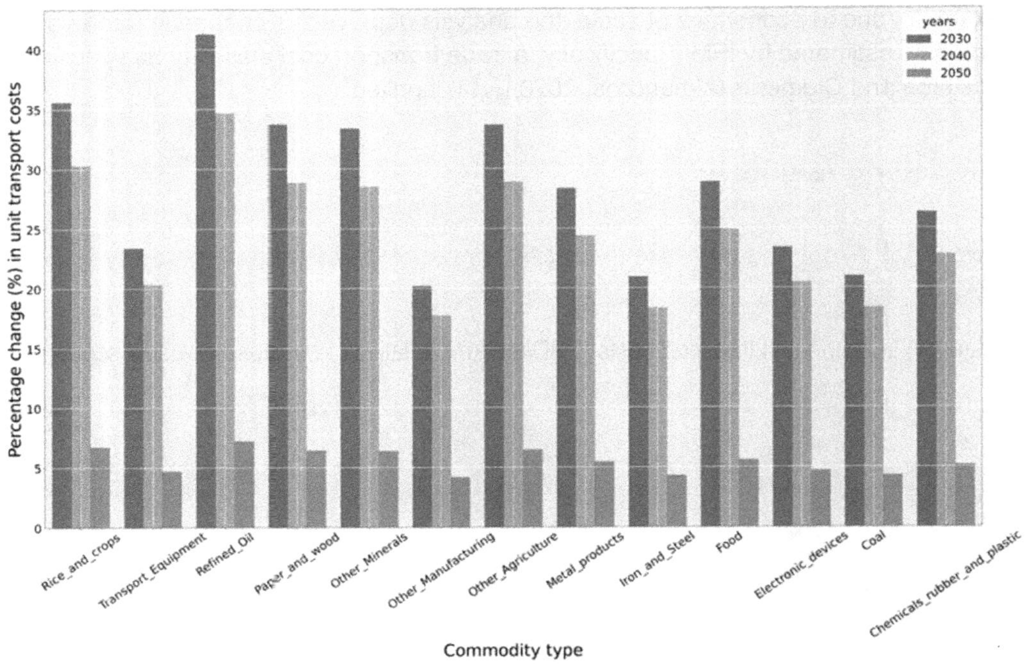

Source: (Equitable Maritime Consulting, 2023[7])

Annex Figure 3.A.3. Projected change in import values from top 22 countries to Hamburg (%)

Percent change (%) in import value (Billion USD), relative to business-as-usual scenario

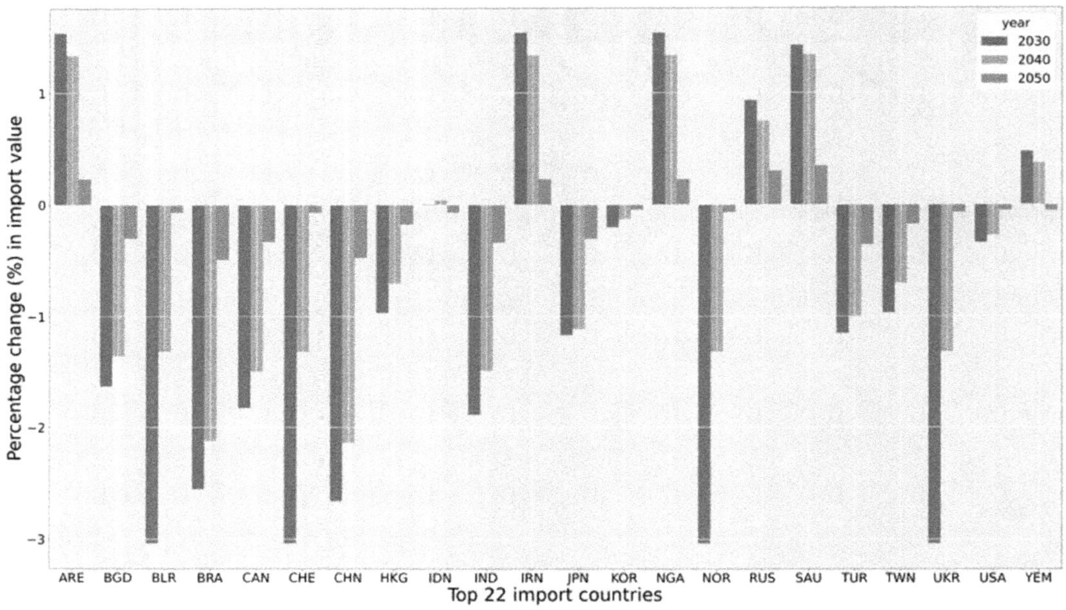

Source: (Equitable Maritime Consulting, 2023[7])

Annex Figure 3.A.4. Projected change in export values from Hamburg to the top 20 countries

Percent change (%) in export value (Billion USD), relative to business-as-usual scenario

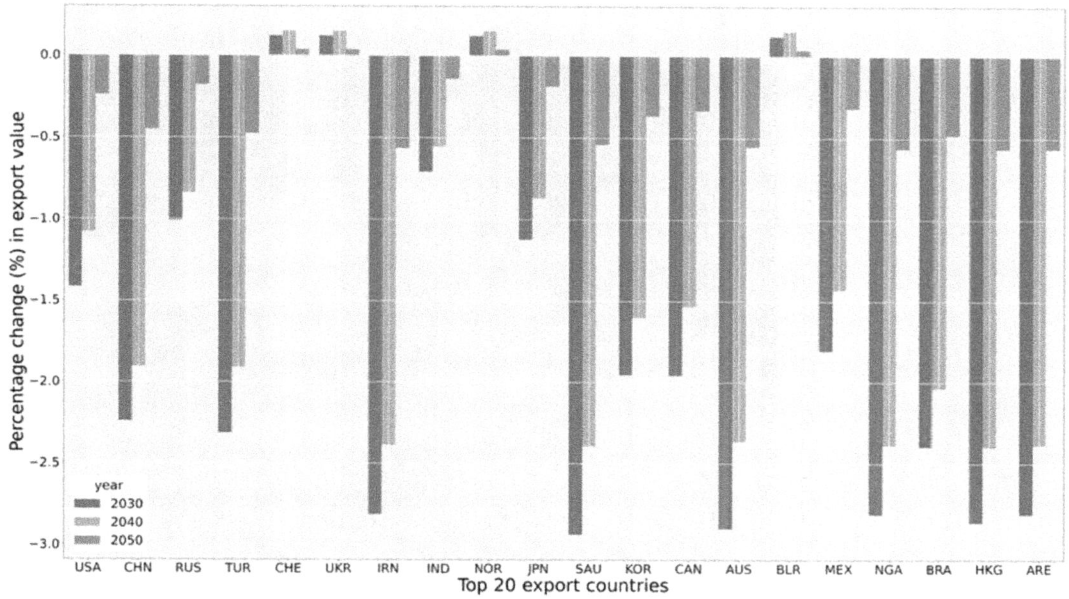

Source: (Equitable Maritime Consulting, 2023[7])

The comparison of different years reveals similarities to the absolute change findings. However, the commodity most significantly affected in terms of percentage is no longer food. Instead, rice and crops experience the most substantial percentage decline in imports, amounting to 4.1% in 2030. Additionally, the export value of refined oil sees a reduction of approximately 8%.

Annex Figure 3.A.5. Projected change in import value of each commodity type to Hamburg (%)

Percent change (%) in import value (Billion USD), relative to business-as-usual scenario

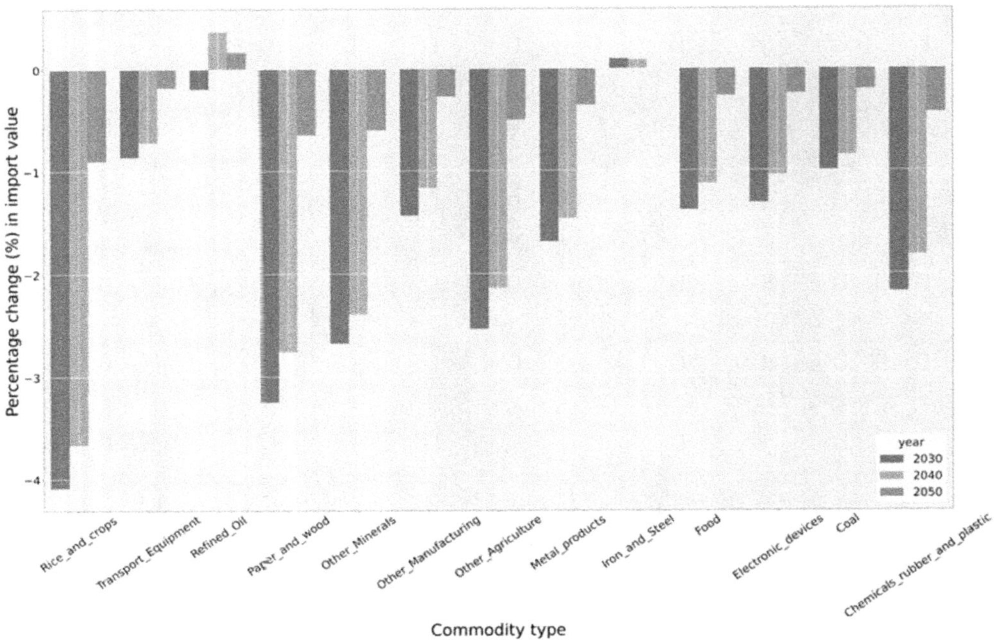

Source: (Equitable Maritime Consulting, 2023[7])

Annex Figure 3.A.6. Projected change in the export value of each commodity type from Hamburg (%)

Percent change (%) in export value (Billion USD), relative to business-as-usual scenario

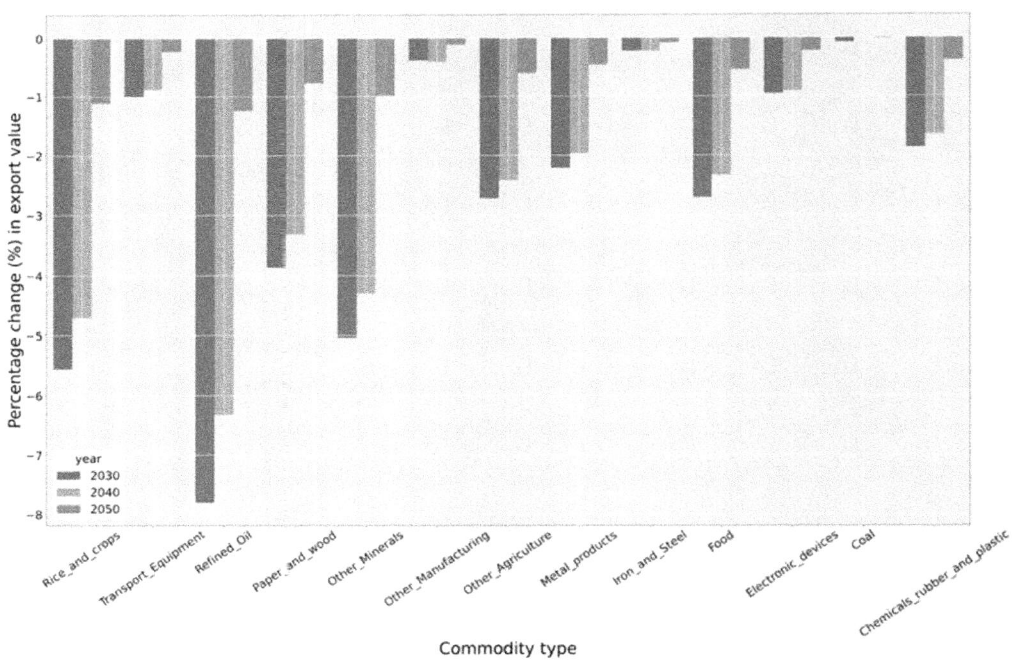

Source: (Equitable Maritime Consulting, 2023[7])

4 The circular economy as a driver for climate neutrality

Chapter 4 first argues that moving towards a circular economy can effectively contribute to achieving climate neutrality in the business sector and shows the role of the business sector in the transition. The chapter then analyses current and planned circular economy initiatives led by the Department of Environment, Climate, Energy and Agriculture of the City of Hamburg (BUKEA). It provides an international literature review on the role of the business sector in promoting the circular economy, highlighting the main barriers to implementation. The chapter concludes by illustrating how the City of Hamburg and the Hamburg Chamber of Commerce can promote the circular economy through new business models.

The transition towards a circular economy as a means to achieve climate neutrality

A circular economy (Box 4.1) can help reduce greenhouse gas emissions (GHG) and contribute to climate neutrality. Material management represents up to two thirds of global greenhouse gas emissions (GHG). It is estimated that in the absence of new policies, GHG emissions related to materials management would rise by 66.6% between 2017 and 2060, from 30 Gt CO_2-equivalents to about 50 Gt CO_2-eq. The transition towards a circular economy could reduce global GHG emissions by 39% by 2032 compared to 2019 levels (OECD, 2019[1]). In cities, the adoption of a circular economy framework in five key areas - namely steel, plastics, aluminium, cement, and food – could achieve a reduction of a total of 9.3 Gt CO_2-eq of GHG by 2050 (Ellen MacArthur Foundation, 2021[2]). It could also do so by substantially lowering energy inputs, a key concern for the transition to climate neutrality. In addition, products designed and manufactured according to circular principles tend to have a smaller environmental footprint than their conventional counterparts.

Box 4.1. Defining the circular economy

More than 100 definitions exist worldwide to characterise a circular economy (Blomsma and Brennan, 2017[3]; CIRAIG, 2015[4]; Homrich et al., 2018[5]). The OECD Expert Group on a new generation of information for resource-efficient and circular economy (RECE-XG) and the UNECE Task Force on measuring the circular economy (UNECE-TF) define a circular economy as one where: i) the value of materials in the economy is maximised and maintained for as long as possible; ii) the input of materials and their consumption is minimised; and iii) the generation of waste is prevented and negative environmental impacts reduced throughout the life-cycle of materials (OECD, 2022[6]). Therefore, moving towards a circular economy goes beyond waste management and recycling; it implies changes in production and consumption models, eco-design and integrated planning. While recycling has an important role to play in the transition, recycling is not as effective as other "Rs" of the circular economy (i.e. Reduce, Reuse, Regenerate, Refurbish, Remanufacture, and Recover).

According to the European Commission, a circular economy is where the value of products, materials and resources is retained in the economy for as long as possible, including by returning them to the product cycle at the end of their use, thus minimising the generation of waste (European Commission, 2015[7]).

The Ellen MacArthur Foundation describes a circular economy as an alternative to a traditional linear economy (make, use, dispose), being restorative and regenerative by design (EMF, 2013[8]).

The International Organisation for Standardisation (ISO) defines the circular economy as an economic system that uses a systemic approach to maintain a circular flow of resources by regenerating, retaining or adding to their value, while contributing to sustainable development (ISO, 2021[9]).

Source: OECD (forthcoming), 21IT01 - Advanced policy instruments to accelerate the circular economy in Italy.

Findings from the OECD (2020[10]) show that climate change is a very relevant driver for the transition to a circular economy for 73% of the 51 cities and regions surveyed[1] (Figure 4.1). Several circular economy strategies at national and subnational level recognise the importance of the circular economy to achieve climate neutrality. For instance, the Spanish Circular Economy Strategy (*España Circular 2030*) sets the goal of achieving climate neutrality by 2050 (Government of Spain, 2020[11]). In Scotland (United Kingdom, UK), *Making Things Last: A Circular Economy Strategy for Scotland* aims to tackle climate change and curb emissions arising from the consumption of goods, as it is estimated that a more circular economy could reduce carbon emissions by 11 million tonnes per year by 2050 (Scottish Government, 2016[12]). London (UK) is pursuing circularity to make a substantial contribution to the mayor's aspiration to achieve a zero-carbon city by 2050 (OECD, 2020[10]). The city of Joensuu (Finland) is planning circular economy actions within its ongoing climate programme that aims for a carbon neutral city by 2025. For the city of Umeå (Sweden), the transition towards a circular economy represents a paradigm to stimulate businesses while achieving the environmental goal of carbon neutrality by 2040, while Glasgow (UK) aims to become the first circular city in Scotland through innovative solutions to achieve carbon neutrality by 2030 (OECD, 2020[13]) (OECD, 2021[14]).

Figure 4.1. Drivers of the circular economy in surveyed cities and regions

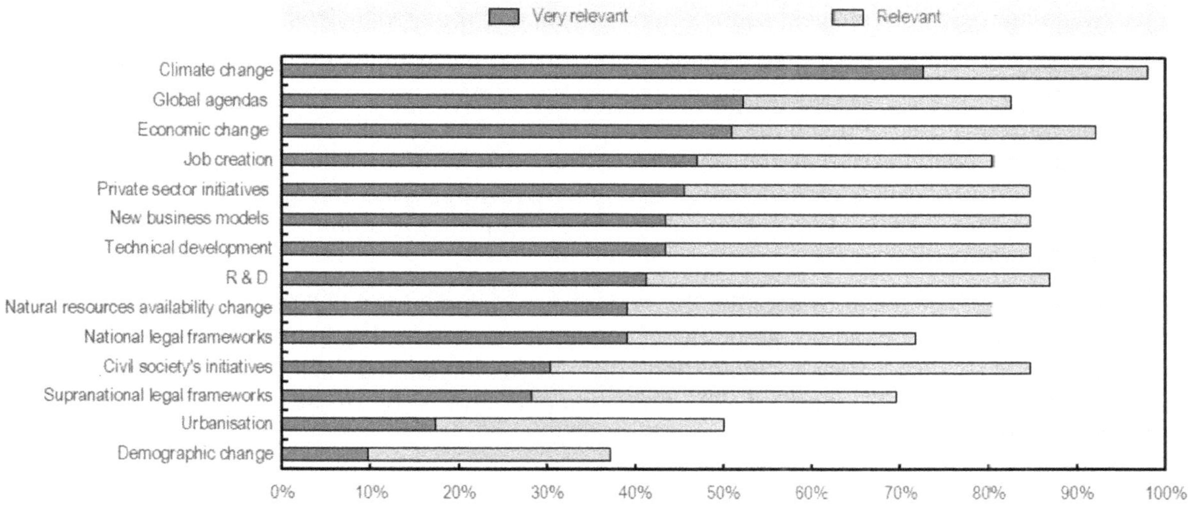

Note: Results based on a sample of 51 responding city representatives.
Source: OECD (2020[10]), *The Circular Economy in Cities and Regions: Synthesis Report, OECD Urban Studies, OECD Publishing, Paris*, https://doi.org/10.1787/10ac6ae4-en.

The circular economy is gaining momentum across several cities, focusing on specific sectors. For example, Rotterdam (Netherlands) focuses on reducing textile waste and increasing recycling by promoting the exchange of used clothing and opening a chemical recycling facility for the treatment of discarded textiles. Seattle (United States) is focusing on the deconstruction of buildings while conserving building materials. Stockholm (Sweden) is developing a digital system to make recycling more accessible to the public. Many German cities are advancing in the transition by incorporating circular economy principles into their waste management plans (e.g. Munich); linking emission reduction targets to the transition to a circular economy (e.g. Berlin, Freiburg, Munich); or implementing circular business models (Aachen, Frankfurt) (Box 4.2).

Box 4.2. German cities' actions to move towards a circular economy

Several German cities are taking steps towards a circular economy:

- In 2021, the city of **Aachen** committed to integrating circular economy principles into administrative processes, development strategies and urban planning, with the participation of citizens.

- In 2016, the city of **Freiburg** joined the EU-Interreg Alpine Space Project Greencycle to reduce CO2 emissions and develop a holistic circular economy system. As part of the project, Freiburg has identified 12 key principles for the transition to a circular economy, including the integration of circular economy principles into local policies, the pursuit of Sustainable Development Goals, the promotion of co-operation, education, sustainable consumption, innovation, infrastructure investment, and monitoring and measurement.
- The **city of Frankfurt** is moving towards a circular economy through an approach based on improving waste prevention and treatment. The city aims to become zero-waste by significantly increasing the share of waste that gets recycled (currently only 45%) and reducing the total waste generated, which is estimated at 280 000 tonnes per year. Initiatives include tackling food waste through participation in the "Cities against Food Waste" network, and supporting the establishment of a Re-Use Network in the State of Hesse to improve co-operation between waste management entities, second-hand shops, and repair initiatives.
- The **city of Munich**'s adopted a resolution in 2020 to promote the implementation of a Circular Economy and a Zero Waste Strategy. The decision emphasises the importance of co-operation and networking among various stakeholders in the city and highlights the need to analyse the status and potential of material flows. Moreover, in 2016, the waste management corporation of Munich (*Abfallwirtschaftsbetrieb München, AWM*) launched the Halle 2 project to create synergies between waste collection and reuse by opening a second-hand store funded by waste collection fees.
- In 2021, the city of **Berlin** adopted the Waste Management Plan (2020-2030), which serves as a planning tool focused on strengthening a circular economy that prioritises waste prevention, reuse and recycling. Between 2020 and 2022, the implementation of the strategy contributed to cutting over 900 000 tonnes of CO_2 emissions and saving approximately 2.2 million tonnes of primary raw materials per year.

Source: OECD (2020[10]), *The Circular Economy in Cities and Regions: Synthesis Report, OECD Urban Studies, OECD Publishing, Paris, https://doi.org/10.1787/10ac6ae4-en;* Circular Cities Declaration (2023[15]), Circular Cities Declaration: Current Signatories, Senate Department for Mobility, Transport, Climate Protection and Environment, (2023[16]), Zero Waste Strategy of the State of Berlin.

Overview of the main ongoing and planned initiatives on the circular economy in Hamburg

In December 2021, the Hamburg Chamber of Commerce (HCC) adopted a resolution committing to building a climate neutral business community in the city by 2040, while the city aims to become carbon-neutral by 2045 as outlined in the 2020 Hamburg Climate Protection Act (*HmbKliSchG*) (City of Hamburg, 2020[17]). The Ecology and Economy Unit within the Ministry of Environment, Climate, Energy and Agriculture (BUKEA) of the city of Hamburg is in charge of exploring opportunities for transitioning from a linear to a circular economy in collaboration with the Waste Policy Unit, based on a legal framework that promotes sustainable waste management and climate policies (Figure 4.2).

Figure 4.2. Timeline of the legal framework for sustainable waste management and climate policies in Hamburg, Germany

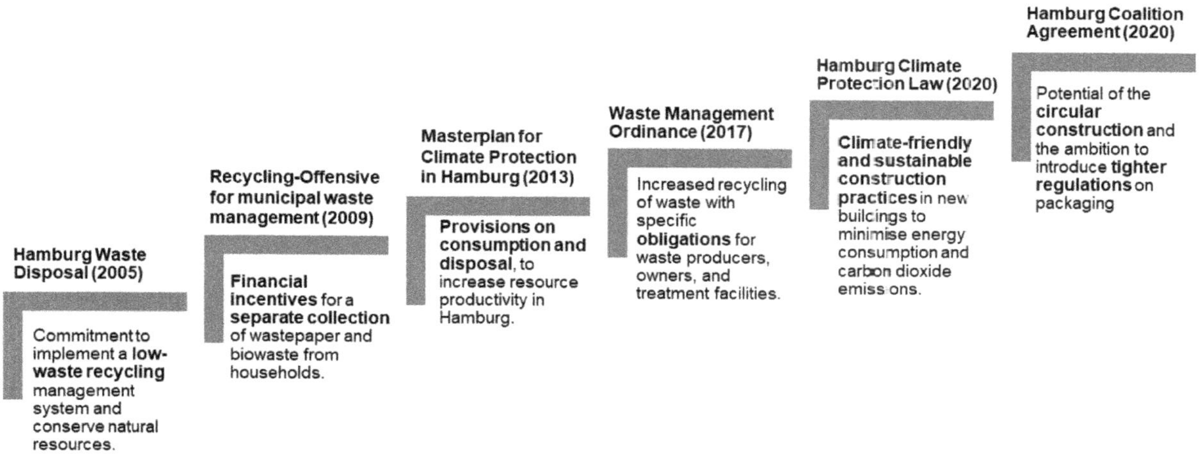

Source: City of Hamburg (2005[18]), Hamburg Waste Disposal Act ; City of Hamburg (2009[19]) Recycling-Offensive für den Klimaschutz; Federal Ministry of Justice (2012[20]), Federal Waste Avoidance, Recovery, and Disposal Act; City of Hamburg (2011[21]), Hamburg Recycling Ordinance; City of Hamburg (2013[22]), Masterplan for Climate Protection in Hamburg; City of Hamburg (2017[23]), Waste Management Ordinance ; City of Hamburg (2020[24]) Hamburg Climate Protection Law; and City of Hamburg (2020[25]), Hamburg Coalition Agreement .

In addition to the regulation listed in Figure 4.2, sectoral waste legislation in Germany and in Hamburg addresses collection, use and treatment of of a range of waste streams, often to prevent specific environmental impacts from poor disposal. This legal framework includes the following: End-of-Life Vehicles Ordinance (BMUV, 2023[26]); Waste Oil Ordinance (BMUV, 2002[27]); Commercial Waste Ordinance (Hamburg CIty, 2017[28]); Batteries Ordinance (BMUV, 2021[29]); Waste Electrical Equipment Ordinance (BMUV, 2022[30]); Waste Wood Ordinance (BMUV, 2023[31]); Sewage Sludge Ordinance (BMUV, 2017[32]); and the Substitute Building Materials Ordinance (BMUV, 2023[33]).

National regulations, which comply with European Union legislation, are also encouraging the adoption of more sustainable materials use to reduce GHG emissions, as well as other environmental impacts, notably from plastics. For instance, the German Federal Government has adopted the Closed Substance Cycle Management Act (KrWG) and the Packaging Act (VerpackG) to implement the 2019 EU Directive on single-use plastics[2], which aims to avoid short-lived plastic packaging, and ban harmful and detachable plastic items. The Single-Use Plastics Prohibition Ordinance (*EWKVerbotsV*) and the Single-Use Plastics Labelling Ordinance (*EWKKennzV*) entered into force in 2021. The former bans the import of plastic products such as disposable cutlery, drinking straws, plates, stirrers, food containers and beverage cups made of expanded polystyrene (e.g. boxes with or without lids). The latter requires the labelling of certain environmentally harmful products, such as disposable plastic cups, cigarette filters, etc. From January 2023, an additional provision of the EU Directive was incorporated into national law, requiring final distributors of single-use packaging to also offer reusable alternatives (City of Hamburg, 2022[34]).

In 2022, the BUKEA commissioned a study on the potential of the circular economy in Hamburg to the Wuppertal Institute and the Institute for Innovation, Climate Protection and Circular Economy (HiiCCE). Preliminary findings of the study suggest the potential role that BUKEA could play, namely acting as a co-ordinating and contact point, a funding authority and a driver for awareness raising. The study also highlights: (i) a need for better co-ordination of all ongoing initiatives; (ii) insufficient dissemination of knowledge on the circular economy; (iii) limited awareness of key stakeholders; (iv) the need to support companies in specific areas, such as supply chain legislation, sustainability reporting and digital material passports[3] (Box 4.3).

Box 4.3. Digital material passports: Tracking materials for better use

Digital material passports provide a comprehensive record of the materials used in the design and manufacture of a product. They aim to enhance transparency and traceability by providing clear information on the origin, composition, and environmental impact of materials via open-source digital platforms. For example, digital passports identify the materials used in buildings during construction and renovation. Among other benefits, they help avoid the costs associated with hazardous materials testing prior to demolition and reuse. They also help make informed decisions about resource management, reuse, recycling, and waste reduction. Some international examples of digital material passports include the following:

- The city of Mikkel (Finland) is applying circular material management methods to carry out the demolition of a healthcare centre and a hospital. Following a selective demolition process, the municipality will digitally track the recovered materials, which will eventually be put to new use on a building materials market.
- The city of Amsterdam (Netherlands) introduced material passports as one of the main action points of its circular economy agenda in 2016. One of the proposed actions was to encourage construction companies to use material passports by offering discounts on land acquisition.

Source: OECD (2023[35]), *The Circular Economy in Tallinn, Estonia*, https://www.oecd.org/publications/the-circular-economy-in-tallinn-estonia-06abc3de-en.htm; OECD (2020[36]), *The Circular Economy in Groningen, the Netherlands, OECD Urban Studies, OECD Publishing, Paris*, https://doi.org/10.1787/e53348d4-en; OECD (2020[13]), *The Circular Economy in Umeå, Sweden, OECD Urban Studies, OECD Publishing, Paris*, https://doi.org/10.1787/4ec5dbcd-en.

This chapter highlights five main typologies of activities undertaken in Hamburg to move towards a circular economy:

1. **Business engagement** in 2003, BUKEA set up the **Hamburg Eco-Partnership** (*Umweltpartnerschaft*), a network of 1 600 companies committed to taking voluntary environmental action. The partnership offers free advice to companies on environmental measures. It brings together partners such as the Hamburg Senate, the HCC, the Chamber of Crafts and Trades (*Handwerkskammer Hamburg),* the Hamburg Industrial Association (*Industrieverband Hamburg*) and the Hamburg Port Association (*Unternehmensverband Hafen Hamburg*). The 2023-2028 work programme prioritises the circular economy and is developing a working group on circular textiles (City of Hamburg, 2023[35]). The **Circular Hub North**, supported by the German Federal Foundation for the Environment (*Deutsche Bundesstiftung Umwelt*, DBU), is one of the four circular hubs in Germany to foster knowledge sharing and collaboration across small and medium-sized enterprises (SMEs) at regional level. In 2023, the BUKEA organised a meeting in Hamburg with 60 stakeholders to kick off the project (City of Hamburg, 2023[36]). Finally, the **Caught-up Initiative** (*aufgefangen*) was initiated by the Ministry of Justice and Consumer Protection of Hamburg in 2021 and promotes solutions against food waste, involving supermarkets, grocery shops and restaurants.
2. **Awareness raising and knowledge building** on the opportunities of a circular economy: in 2022, the city hosted the "1st Hamburg Dialogue against Food Waste". The Ministry for Justice and Protection and the Hamburg University gathered experts to discuss practical solutions and share knowledge on best practices to reduce food waste (City of Hamburg, 2022[37]). In the fashion industry, the city organises 'Green Fashion Tours", where participants have the opportunity to hear and learn from local designers and store managers about how they incorporate sustainable and circular economy principles into their businesses, as well as the selection of materials, suppliers

and products and the design of second-hand products (Hamburg Tourism Office, 2023[38]). In June 2023, the city of Hamburg published its first **Sustainability Report**. According to the report and in relation to SDG 12 on sustainable production and consumption, the city applies a five-stage waste hierarchy, as an obligation of the KrWG, with waste prevention as the highest priority, followed by reuse, recycling, energy recovery and disposal. Between 2015 and 2020, the amount of waste per inhabitant decreased from 0.45 to 0.44 tonne, reflecting some progress in complying with this waste hierarchy (City of Hamburg, 2023[39]). BUKEA plans to develop a Circular Economy Strategy for Hamburg with a focus on business development and concrete measures to build knowledge around the circular economy in Hamburg. Launched in 2023, the **Hamburg Port Development Plan 2040** also includes measures to further strengthen and expand the circular economy in the port and to increase resource efficiency, mainly in the built environment and infrastructure. Some of the actions included in the plan are: site remediation and activation of contaminated soil; increasing space efficiency in buildings through multi-storey designs; cooperation with the city of Hamburg in researching suitable areas for soil storage; regular use of old asphalt in the construction of new roads; and use of waste from port operations to supply heat (Hamburg City, 2023[40]).

3. **Networks**: In 2019, Hamburg became the first **German Fab City**, along with 41 cities around the world. The international Fab City Foundation has set the goal of completely transforming the urban economy of Fab Cities into a circular economy by 2054, where no physical goods or raw materials are imported or exported and where all products consumed are produced locally. Funded by the Ministry for Economy and Innovation, the Fab City Hamburg project aims to promote science and research, environmental protection, public education and vocational training, as well as art and culture. As part of the Fab City Hamburg project, the BUKEA organises awareness-raising workshops (Fab City Hamburg, 2022[41]). Hamburg participates in the the project **Circular City – Opportunities for local and regional resilience & value creation** ("*Kreislaufstadt – Chancen für Resilienz und Wertschöpfung*") led by Difu (German Institute of Urban Affairs). The Difu aims to support municipalities in developing their own city-wide circular economy strategy based on the political and legal framework and on the knowledge of already active circular economy cities and initiatives. The project will take place from July 2023 to February 2025 (Difu, 2023[42]).

4. **Guidelines:** Hamburg's website provides clear **guidelines** on how residents can donate used clothes to the Hamburg City Cleaning Service (SRH), the city's largest waste collection and treatment service provider. The SRH facilities include two second-hand department stores and 12 recycling centres where citizens can directly dispose of recyclable used textiles in small quantities of up to 1 m^3 free of charge. Rags and leftovers from pre-sorting are accepted as residual waste for a fee. Hamburg provides information on prices and fees for commercial customers and a list of collection points for used textiles in the city. In addition, the Hamburg Consumer Advice Centre offers information and consultation services to consumers to enhance market transparency. On its website, it features a list of sites where residents can donate discarded clothing (Hamburg Consumer Advice Centre, 2022[43]). The city website provides information on rules and legal frameworks applying to warehouses (*Fairteilers*), where both households and businesses can donate edible food. The donated food is stored in refrigerators and on shelves, and is made available to the local community free of charge (City of Hamburg, 2023[44]). Moreover, the Hamburg Tourism Office (*Hamburg Tourismus*) promotes sustainable tourism practices by advertising green transport and accommodation, as well as sustainable activities, to visitors. Since 2018, the tourism office has received the Green Globe certification, recognising its efforts to reduce resource consumption, minimise waste, use seasonal ingredients, increase recycling, share resources, and offset GHG emissions (Hamburg Tourismus, 2023[45]). The Hamburg Tourism Office offers guidelines and advice for tourists to spend a sustainable holiday in the city and provides a list of sustainable hotels, which are sorted by different criteria such as avoiding using plastic (Hamburg Tourism Office, 2023[46]). The Tourism office highlights the importance of adopting a sustainable tourism policy in the city, as recognised in the city website with recommendations aiming to use

fewer resources and energy, produce less waste and increasing the recycling (Hamburg Tourism Office, 2023[47]).

5. **Funding**: in co-operation with the Hamburg Investment and Development Bank (*Hamburgische Investitions- und Förderbank*, IFB Hamburg), the city of Hamburg has implemented two funding schemes to support the transition towards a circular economy: (i) UfR - companies for resource protection (*Unternehmen für Ressourcenschutz*), and (ii) the Programm für Innovation (Programme for Innovation, PROFI) Umwelt and PROFI Umwelt Transfer schemes. The former, in place since 2001, aims to encourage investment in increasing energy and resource efficiency in the operational processes of companies. The programme proposes loans and grants from EUR 1 000 up to EUR 100 000 for individual projects, depending on the amount per tonne of CO_2 avoided annually or per tonne of material, waste or cubic metre of water saved. The PROFI Umwelt and PROFI Umwelt Transfer schemes, within the framework of the PROFI Programme, promote individual and collaborative projects that foster the development of innovative products, processes and services that help save resources and emissions. The schemes provide up to EUR 500 000 for individual projects and EUR 1 million for collaborative projects from companies of all sizes, sectors and technologies, as well as universities and research institutions co-operating with them (IFB Hamburg, 2023[48]; 2023[49]). Companies and research institutions also receive subsidies for industrial research, experimental development, and feasibility studies. In addition, Hamburg has joined three major EU funded programmes aiming to support the transition to a circular economy. First, in 2016 the city joined the **European research project "FORCE"**, which aimed at demonstrating the potential of circular practices in relation to the waste prevention and treatment for four key materials (plastics, electric waste, wood waste, and biowaste). As a result of this project, Hamburg launched a tool to inform citizens about reselling, repairing, recycling and donation possibilities of Waste from Electrical and Electronic Equipment (WEEE) (Hamburg City, 2016[50]; EC, 2022[51]). Second, the city of Hamburg, together with the cities of Helsinki (Finland), Copenhagen (Denmark) and London (UK), joined the **Circular Construction in Regenerative Cities (CIRCuIT) project**, which is funded by the EU's Horizon 2020 programme between 2019 and 2023. The main objective was to close the gap between theory and implementation in the adoption of circular economy principles in the built environment sector. CIRCuIT established a knowledge exchange structure to enable the up-scaling and replication of circular economy solutions (CIRCuIT, 2023[52]; Hamburg City, 2019[53]). Third, the City of Hamburg and HafenCity University participated in the European Union Horizon 2020-funded **REPAiR project** in 2016, which aimed to provide local and regional authorities with an innovative transdisciplinary open source Geodesign Decision Support Environment (GDSE). The overall objective of the project was to develop spatial sustainable development strategies, while demonstrating the feasibility and validity of the GDSE as a tool to improve waste and resource management (Hamburg City, 2016[54]; REPAiR, 2023[55]).

The role of the business sector

Introduction

The circular economy is a shared responsibility across levels of government and key stakeholders including the business sector. A total of 80% of cities and regions responding to an OECD survey (2020[10]) have identified the business sector as a key player in contributing to the development and implementation of circular economy initiatives (Figure 4.3). Businesses and other economic actors are generally involved in the development of circular economy strategies, as the private sector is leading the shift towards new business models. For example, the Dutch, Finnish and Slovenian roadmaps recognise the key role of SMEs in the transition to a circular economy. In Greece, the inclusion of the circular economy into the national growth strategy involves initiatives aimed at building knowledge, as well as linking entrepreneurship and social economy with technological innovation (OECD, 2020[10]).

Figure 4.3. Type of stakeholders involved in the development of circular economy initiatives in surveyed cities and regions

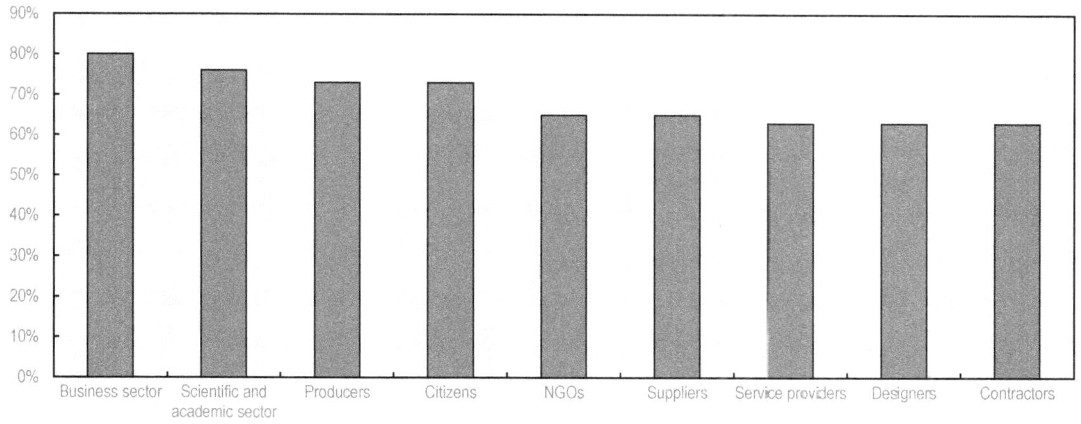

Note: Results based on the 51 surveyed cities and regions that responded "Yes" to the question: "Have the following categories of stakeholders been involved (or planned to be) in the design and implementation of the circular economy initiative of your city/region?".
Source: OECD (2020[10]), *The Circular Economy in Cities and Regions: Synthesis Report, OECD Urban Studies, OECD Publishing, Paris*, https://doi.org/10.1787/10ac6ae4-en.

New business opportunities and the development of new markets represent an important driver for cities and regions to transition to a circular economy. A total of 85% of respondents to the OECD survey consider the emergence of new business models as a very relevant or relevant driver (Figure 4.1). While there is no consensus on the exact definition of a circular business model, it is usually presented as an "alternative value proposition" through which companies can reduce their environmental footprint by creating and capturing the value of a given material input (Ekins, P. et al., 2019[56]). Lahti et al. (2018[57]) argue that "*a circular business model is designed to create and capture value while helping achieve an ideal state of resource use (e.g. finding a model that most closely resembles nature and comes close to achieve complete material cycles*)". Implementing circular business models also has a positive socio-economic impact. In 2019, circular activities (e.g. repair, reuse or recycling) generated around EUR 145 billion in value-added in the EU-28, 30% more than in 2012 (Eurostat, 2022[58]; Eurostat, 2022[59]). According to the European Commission, the share of employment in the circular economy increased by 13% between 2012 and 2019.

All sectors are concerned, but some show higher potential in terms of economic, social and environmental benefits, including the reduction of GHG emissions. Making a sector "circular" implies rethinking value chains, production and consumption processes. "Circularity" entails that any output can be an input for something else within and across sectors. It aims to: make products and goods last longer through better design; produce goods using secondary and reusable materials, and renewable energy, while reducing atmospheric emissions; produce and distribute products locally and consume them in a conscious and sustainable manner; and transform waste into a resource (OECD, 2020[10]).

A number of examples of circular business models described below concern the retail and the hospitality sectors[4] in Hamburg. These sectors hold potential in developing circular practices towards a more efficient use of resources. The retail industry covers all activities of reselling new and second-hand goods (excluding motor vehicles and motorcycles), mainly to the general public for personal or household consumption or use. The retail industry plays a pivotal role in OECD countries, as it links upstream industries and consumers, contributes almost 5% of GDP and employs around 1 in 12 workers (OECD, 2020[60]). In addition, before the COVID-19 pandemic, in 2019 (or latest available year pre-COVID-19) the tourism and hospitality sector contributed 4.4% of GDP in OECD countries, generates 21% of service exports and accounts for about 6.9% of total employment (OECD, 2022[61]). In Germany, between 2013 and 2021, the retail sector reduced its CO_2 emissions by 33% (National Climate Initiative, 2023[62]).

Circular business models can be classified according to a value chain approach, which divides business models into circular design, optimal use, and value recovery; or they can be based on the material flow they address. The OECD (2019[56]) identifies five types of circular business models, which are described below, namely: Circular supply models; Resource recovery models; Product life extension models; Sharing models; and Product service system models (PSS). In practice, firms tend to combine several business models rather than following a specific one. For instance, adopting a circular supply model, where strategic sourcing and design decisions are made early in a product's life, can improve the business case for component and material recovery further downstream (OECD, 2019[63]).

Circular supply models replace conventional material inputs obtained from virgin resources with bio-based, renewable, or recovered materials, which reduces the need to extract virgin resources. This is the main type of circular business model implemented among respondents to a business survey on climate neutrality targets and actions among HCC membersfollowed by the share of capital goods (29%) and refurbishment and remanufacturing (29%) (Figure 4.4). Responses may not be representatives as only a fraction of businesses responded and businesses may be more likely to respond if they have stronger interest in environmentally-friendly business practices. Circular supply models allow firms to market their products as "green" and target environmentally conscious consumers. They can also help manage supply chain risks, as the raw materials used to produce key inputs are often located in a limited number of countries. Circular supply models also require increased co-ordination within and across the value chain (OECD, 2019[63]) (Box 4.4).

Box 4.4. International examples of cities implementing circular supply models

- In 2018, Amsterdam (The Netherlands) established the Circular Hotels Leaders Group (*Kloplopergroep*), comprising 12 hotels actively incorporating circular principles into their business models. This collaboration included knowledge sharing, joint purchasing and consolidation of waste streams, reuse of furniture and carpet tiles, bed repair and reuse, introduction of *a la carte* breakfast options, and collaborative circular procurement practices with other hotels, including sustainable linen and laundry rental services.
- The Network for Sustainable Construction and Real Estate Management in Cold Climates (*Nätverket för hållbart byggande och förvaltande*) launched by the city of Umeå (Sweden) in 2008 brings together 55 members from all segments of the construction supply chain. Sustainability and the circular economy are at the core of the network's annual member meeting and monthly breakfast meetings. The network has enabled the creation of a public-private partnership (PPP) to develop, by 2024, the new Tomtebo Strand city district, which incorporates circular economy principles in its structural plan.
- The Sustainable Restaurants Network (*Hållbara Restauranger*) involves 14 restaurants in Umeå (Sweden) for sustainable practices in the food industry and food waste management. On a wider scale, the North Sweden Cleantech is a regional innovation platform focusing on exporting green technology, clean energy and sustainable solutions through business support and networking. A hundred companies are currently part of the platform. Since 2016, the platform has been organising circular economy capacity building events.
- In Ireland, the National Platform for Circular Manufacturing Initiative 2020-22 (CIRCULEIRE) is the first cross-sectoral industry-led innovation network dedicated to accelerating a zero-carbon circular economy. It is a PPP de-risking and delivering circular business model innovation. The platform is led by Irish Manufacturing Research (IMR), in collaboration with its strategic partners, the Department of Communications, Climate Action and Environment (DCCAE), the Irish Environmental Protection Agency (EPA), EIT Climate-KIC and 25 industry members. Companies involved come from the building, furniture, packaging and material reprocessing sectors, among others. The platform has a dedicated innovation fund designed to foster cross-sectoral systems integration projects.
- Paris's (France) 1st Circular Economy Roadmap has set the objective of promoting sustainable, organic and responsible product supplies in public entity canteens (e.g. in schools). This objective is linked to the implementation of a more socially and environmentally responsible public procurement scheme and the goal of expanding urban agriculture practices in the city.
- Rotterdam (The Netherlands) is using a digital market place for building materials, bringing supply and demand of used building components together and promoting sustainable supply in the construction sector.
- By including the circular economy in the textile supply chain and the reuse and recycling of clothing, London (UK) aims at: reducing the number of textiles sent for disposal; becoming a hub for textile collection, reuse and recycling; and being recognised as a well-known circular economy textile design centre. In 2017, 330 000 tonnes of textiles were diverted from landfill in the UK.

Source: Network for Sustainable Construction and Real Estate Management in Cold Climates (2013[66]), Umeå. More Sustainable Buildings; Municipality of Umeå (2019[67]), Tomtebo Beach - A New Neighborhood with People and Sustainability in Focus; North Sweden Cleantech (2019[68]), North Sweden Cleantech; Irish Manufacturing Research (2020[69]), CIRCULÉIRE – The National Platform for Circular Manufacturing, and Traid (2018[70]), The Impacts of Clothing .

Figure 4.4. Circular business models in Hamburg, Germany

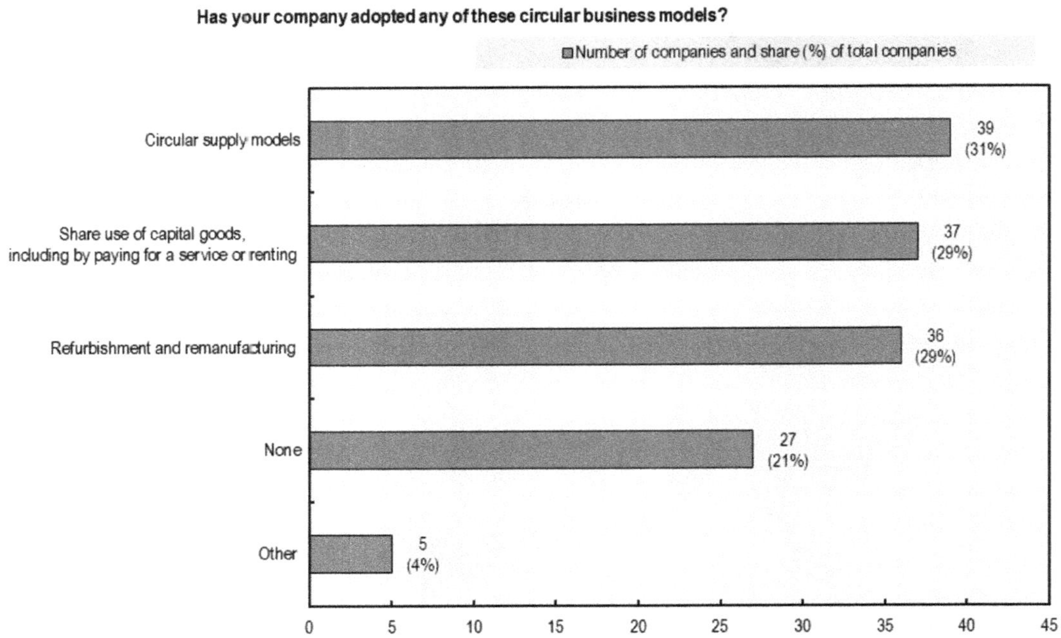

Note: The survey was sent to a sample that is broadly representative of the population of Hamburg businesses in the sectors covered by the HCC, although businesses with fewer than 10 employees are underrepresented. The questionnaire was sent to a random sample of 2300 businesses with more than 10 employees and to a random sample of 500 businesses of all size classes. 128 businesses responded, a response rate of about 4%.
Source: Survey carried out by HCC (2023).

In Hamburg, numerous companies have adopted practices that endorse circular models. In the textile industry, these practices primarily focus on the reuse and repurposing of materials, extending the lifespan of used garments through upcycling while diverting them from landfill disposal. The hospitality sector has also taken steps to reduce food waste through an effective management and measurement system. In addition, in the retail sector, companies of the supply chain achieved the resource recovery of plastic packaging waste.

A Hamburg-based fashion start-up called Bridge&Tunnel produces new clothes, accessories, and home items from a variety of used textiles. The materials are sourced through collaborations with clothing chambers, private individuals, or surplus production from fashion companies. In addition to the environmental benefits of avoiding materials in the production of new products, the company also provides employment opportunities for socially disadvantaged individuals and refugees (Bridge&Tunnel, 2023[69]). Another example in the fashion industry is the start-up NONOI, which recycles old clothes made of high-quality materials such as cashmere, linen, or wool to create new pieces through upcycling (NONOI, 2023[70]).

In 2016, the Hamburg-based cruise line TUI Cruises, together with Futouris and supported by the United Against Waste Association (UaW), launched a project to reduce food waste on cruise ships based on a user-oriented service system. TUI Cruises and UaW measured food waste stemming from overproduction (return from buffet) and quantified plate leftovers. The project also measured waste from food storage and production waste on the entire ship. Analysis results showed that 50% of TUI Cruises' food waste was generated in the overproduction segment, followed by the segment of plate leftovers (18%). Following this initiative, TUI Cruises cut food waste by 17% between 2016 and 2017 (TUI Cruises, 2019[71]).

In 2020, the city administration and the Hamburg University of Technology (TUHH), along with other stakeholders from the private sector, launched the *Hamburg's Recyclables Initiative* to demonstrate how incentivising recycling at the local level can contribute to closing loops. As part of this initiative, the "Hamburg's Bottle" project aims to create a bottle entirely from recycled plastic waste by allocating each partner a specific role: the Hamburg City Cleaning Service (*Stadtreinigung Hamburg - SRH*) collects plastic waste, while key partners from the private sector are responsible for extracting high-density polyethylene and transforming it into recyclable material, as well as filling bottles with an established detergent. The local drugstore retailer Iwan Budnikowsky GmbH & Co. KG ensures the distribution of these products to consumers in Hamburg. In addition to increasing recycling and reducing the consumption of new materials, the project aims to raise awareness among customers, by concretely demonstrating recycling principles and highlighting their own influence to support regional circular cycles through their purchasing decisions (Interreg Europe, 2021[72]).

Resource recovery models transform waste into secondary raw materials, diverting waste from disposal and displacing the extraction and processing of virgin natural resources. This business model comprises three variants: downcycling, upcycling and industrial symbiosis. *Downcycling* involves transforming waste into secondary materials. *Upcycling* refers to converting waste into secondary materials that are subsequently used in valuable applications. *Industrial symbiosis* refers to a situation where one firm uses the by-products of another company's production as inputs (OECD, 2019[63]). Among the cities hosting industrial symbiosis processes and clusters, the city of Kalundborg (Denmark) fosters eco-innovation amongst eight public and private companies to reuse water and energy and recycle materials. The industrial symbiosis in Drummond (Canada) is a network of local companies exchanging resources, such as waste materials, by-products, equipment, space or even energy. Some companies participating in industrial symbiosis sell their production waste rather than paying to dispose of it, thus making a double economic profit. The Eco Parks in Kitakyushu (Japan) allow for recycling waste while producing energy, saving water and creating new business opportunities. The Metropolitan Project of Industrial Symbiosis in the Barcelona Metropolitan Area (Spain) co-ordinates industrial symbiosis projects with circular economy initiatives (OECD, 2020[10]). In Sweden, the roadmap for industrial symbiosis connects with urban symbiosis. While industrial symbiosis allows resource exchanges across companies, urban symbiosis looks at mutual and beneficial exchanges of resources within urban areas and across industries.

Some companies located in Hamburg have adopted resource recovery models such as Cirplus, a Hamburg-based B2B marketplace for recycled plastics. The platform aims to digitise and shorten the transaction process for plastics processors and recycling companies. Its main objective is to consolidate all aspects of the transaction process including finding, negotiating, contracting, shipping, insuring, and paying for recycled and plastic waste, in order to reduce the cost of recycled plastics compared to virgin plastics (Cirplus, 2023[73]). A company called Recyclehero uses electric cargo bikes across the city of Hamburg to collect used glass, paper, clothing, and deposit bottles from companies and private households destined for recycling. The company is responsible for collecting bottles and boxes and disposing of them in the correct container, while offering customers a subscription option for the regular collection of recyclable materials. In addition, Recyclehero's recruitment process gives priority to people in vulnerable situations, such as the long-term unemployed (Recyclehero, 2023[74]).

Product life extension models prolong the useful life of products, slow down the flow of constituent materials through the economy, and reduce the pace of resource extraction and waste generation. The life of products can be further extended through maintenance; reuse and repair; or refurbishment and remanufacturing processes to take advantage of the cost savings of using pre-existing materials and products as inputs. For example, the city of Tokyo (Japan) has rented equipment for the celebration of the 2020 Olympic Games and leased it after the games. The circular economy will also play a role in the upcoming Paris 2024 Olympic Games through leasing of materials (OECD, 2021[75]). The city of Paris (France) emphasises the need to facilitate the extension of product life cycles and has implemented measures to recover information technology (IT) and telephone equipment and furniture. It has also promoted the adoption of a charter in cultural venues for the design of eco-responsible events (OECD, 2020[10]). At national level, Italy's circular economy strategic framework focuses on product eco-design whereby material resources are rationalised by replacing non-renewable materials with renewable, recycled, permanent, biodegradable, non-hazardous and compostable materials; and recreating production processes to make more products that can be easily disassembled, recycled, modular (replacement of parts, recovery and reuse of systems and sub-systems) and repairable. Similar emphasis is placed on efficient and circular product design in Portugal's circular economy action plan, with a particular focus on boosting manufacturer innovation and responsibility in order to manufacture products that are "designed to last" by encouraging strategies to extend product working life and support the development of a network of repair facilities by establishing partnerships with municipalities to train and disseminate repair and reuse networks.

In Hamburg, a number of supermarket chains, restaurants, businesses and start-ups offer reusable cups and packaging for beverages and food through deposit systems, and the city administration has played a leading role in their implementation. Since 2016, the city's Environment Agency restaurant has only served takeaway coffee in its own reusable porcelain cups. Similarly, a number of cafeterias in local public institutions have also adopted this practice. In 2017, the city of Hamburg launched the "Return.Again" (*Kehr.Wieder)* campaign in collaboration with local coffee shops. As part of the campaign, more than 260 establishments offered discounts of between EUR 0.10 to EUR 0.30 to customers who brought their own reusable cups. An interactive map on Hamburg's official website showed the location of coffee shops, bakeries, and bistros where customers could benefit from a discount. From 2017 to 2018, the campaign helped avoid the use of approximately 120 000 disposable cups (City of Hamburg, 2018[76]). In 2018, the City joined another deposit and return scheme implemented by the start-up RECUP. In this system, customers buy takeaway coffee in a returnable cup, pay a 1 EUR deposit on the cup, and receive an additional price advantage over coffee sold in disposable cups. These cups can be returned to any of RECUP's partner institutions in Hamburg and Germany. In 2018, the City health authority and the Hamburg police academy also introduced the RECUP system in their canteens (City of Hamburg, 2018[76]).

Some companies in Hamburg have experienced high return rates for reusable cups, reaching up to 99.5% in some cases (Hamburg City, 2022[77]) (German Retail Association, 2022[78]). At the national level, the German Retail Association is also active in the development of reusable solutions by supporting small and medium-sized retailers in engaging in this transition (Box 4.5).

Box 4.5. The German Retail Association for a plastic-free retail industry

The German Retail Association (HDE) aims to strengthen climate protection in the retail sector and to offer practical support towards climate-neutrality at the national and federal level. To achieve this goal, the HDE developed the Climate Protection Initiative (*Klimaschutzoffensive* - KSO) in 2017 to support small and medium-sized retailers in developing energy-saving solutions and other measures for climate protection. The Climate Protection Initiative is part of the National Climate Initiative and has been partially funded by the German Federal Ministry for Economic Affairs and Climate Action (BMWK) since 2017. The German Retail Association estimates that this initiative could reduce emissions by 30 000 tonnes of CO_2 within the industry.

Additionally, in 2023, the KSO published a guide entitled "*Reusable instead of more waste*" (*Mehrweg statt mehr Müll*) on the use of reusable packaging in the food retail sector. This document presents a comprehensive overview of the current legal framework conditions and hygiene requirements. It also provides information on why reusable solutions are more sustainable than plastic packaging, how individual reusable pool systems work, and how retail employees can properly communicate the benefits of reusable containers to customers at the point of sale.

In 2017, the German Packaging Act (*Verpackungsgesetz* - VerpackG) made it compulsory for restaurants and retailers to offer reusable alternatives for the delivery of food and beverages. In order to comply with this regulation, the Climate Protection Initiative has designed posters for retailers to display at the point of sale to encourage customers to use reusable products.

Source: New guide to reusable food retailing, https://www.hde-klimaschutzoffensive.de/de/energie-sparen/neuer-leitfaden-zu-mehrweg-im-lebensmitteleinzelhandel [accessed on 2 October 2023].

Within the product-life extension model, maintenance, repair, and refurbishment services are also provided at Repair Cafés in Hamburg. The purpose of these shops is to restore damaged products (e.g. smartphones, flat tires, bicycles, televisions), either for a fee or on a voluntary basis, and to spread knowledge about how to repair items while protecting the environment and promoting a sustainable lifestyle. The official website of the city provides information on the location, schedule and contact details of 13 Repair Cafés based in Hamburg. For instance, the Repair Café Altona has set up a workshop on the repair of electronic equipment. Experts on electronics meet every month to screw, tinker and solder deteriorated electrical appliances, and also act as a network to seek synergies (Hamburg City, 2023[79]).

Sharing models (also named sharing economy or sharing platform) facilitate the sharing of under-utilised products while reducing the need for new products and the raw materials they contain. This model is particularly suited to densely populated urban areas to reduce the transaction costs associated with a temporary change in product ownership (OECD, 2019[63]). Several cities (e.g. Antwerp, Belgium; Lappeenranta, Finland; Lisbon, Portugal; Malmö, Sweden; Milan, Italy; and Paris, France) offer examples of local government support for circular initiatives, notably by investing in circular transport measures with shared municipal car and bicycle fleets, developing areas for the management of logistics activities, enhancing public transport usage, expanding sustainable transportation choices, and establishing additional cycling lanes (OECD, 2020[80]). In Espoo and Helsinki (Finland), as part of the Cycling Promotion Programme 2013-2024, the Urban Environment Sector and the Helsinki Metropolitan Area Transport Ltd. offer a joint bike service that can be used in the city all year long and allows to move between cities during the city bike season from April to October (City of Espoo, 2013[81]).

Product service system models (PSS) involve marketing services instead of products to enhance incentives for green product design and the efficient use of products. These models facilitate an efficient use of natural resources and can be divided into three main variants: product-oriented, user-oriented, and result-oriented PSS models. *Product-oriented models* focus on the product. In this case, manufacture firms continue to produce and sell products in a conventional way but add after-sales services. *User-oriented models* give access to the services associated with a specific good without ownership of the good itself. *Result oriented-models* enable firms to market a service or outcome instead of a product (OECD, 2019[63]). Several cities have put these models into practice. For example, the municipality of Bollnäs (Sweden) has applied "functional public procurement" (*funktionsupphandlingen*) to rent light as a service in municipal pre-schools and schools. The service is provided by a start-up that received support from Umeå's BIC Factory business incubator. The Amsterdam Airport Schiphol (The Netherlands) also rents light as a service, instead of the traditional model of buying light bulbs: Schiphol pays for the light it uses, while Philips, the provider, is the owner of all installations and is responsible for performance and durability, improving its incentives to use cost-effective, long-lasting products.

Main obstacles to the implementation of circular business models

The literature review conducted for this chapter shows a distinction between internal and external barriers from the business perspective. Internal barriers pertain to decisions and strategies made within a company or by individuals, which are, to some extent, under their control. External barriers are shaped by the operating context of the company or individual and beyond the latter's direct control, making it necessary for policy intervention to address them (Figure 4.5). For example, external factors include challenges such as "inconsistent policies and messages" and "lack of clear pricing signals". In addition, barriers pertaining to "supply chain constraints and thresholds in technologies and infrastructure capacity" may involve interactions with other companies.

Internal barriers can be addressed and improved by the company itself. The barriers encompass elements such as the "business and commercial model," "knowledge and expertise", "competing priorities", "internal capacity and resources", "habitual behaviour", and "negative attitudes and cultures". In some cases, external barriers can potentially hinder internal dynamic forces of improvement. For instance, a lack of consumer demand for recycled or remanufactured products, or the absence of adequate policies, can significantly restrain the motivation of a company to remove internal barriers. Similarly, financial barriers including "high upfront costs", "low returns on investment", and constrained "access to capital" reflect internal company conditions but are strongly influenced by external circumstances (Ekins, P. et al., 2019[56]).

Figure 4.5. Barriers to business becoming more resource-efficient

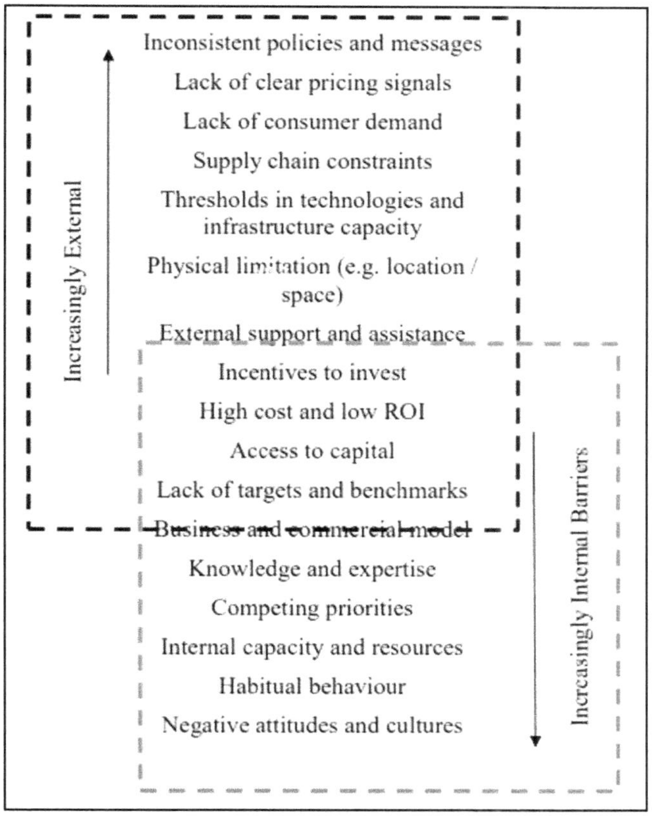

Source: (Ekins, P. et al., 2019[56]) *The Circular Economy: What, Why, How and Where*, https://www.oecd.org/cfe/regionaldevelopment/Ekins-2019-Circular-Economy-What-Why-How-Where.pdf

When implementing circular business models at the subnational level, a number of barriers arise, in particular: lack of awareness, insufficient financial incentives, small scale of circular business models and inadequate regulatory frameworks. Results from the HCC business survey with 126 responding businesses suggest that the main obstacles in the city for the implementation of circular business models relate to: the lack of technological solutions (22% of respondents); lack of conducive regulatory frameworks (14%); lack of awareness of circular business models (13%); lack of skills (12%); lack of critical mass (10%); lack of financial resources (10%); lack of human resources (7%); cultural barriers (6%); and transaction costs (4%) (Figure 4.6).

Figure 4.6. Main obstacles hindering the implementation of circular business models in companies in Hamburg

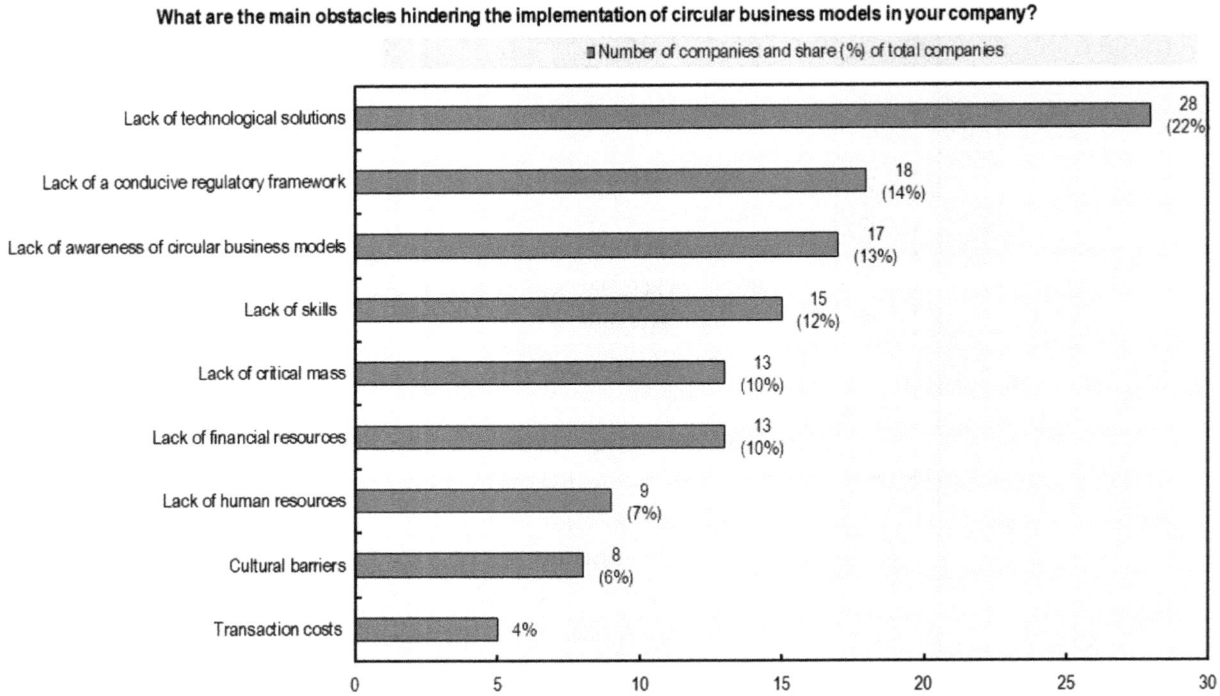

Note: The survey was sent to a sample that is broadly representative of the population of Hamburg businesses in the sectors covered by the HCC, although businesses with fewer than 10 employees are underrepresented. The questionnaire was sent to a random sample of 2300 businesses with more than 10 employees and to a random sample of 500 businesses of all size classes. 128 businesses responded, a response rate of about 4%.
Source: Survey carried out by HCC (2023).

Regulation is not yet conducive to circular-related innovations. At the local level, various regulations are promoting efficient waste management and addressing climate change, but there is no strategic document dedicated to the circular economy.

Although at national level a Circular Economy Act has been adopted and the building blocks of the future national circular economy strategy (planned for 2024) have been identified, at local level there is not yet clear vision of the city's rationale for moving towards a circular economy. In Germany, the Circular Economy Act (2012[82]) promotes the principles of the circular economy and the environmentally friendly management of waste, including the conservation of natural resources, the protection of human health and the preservation of the environment throughout waste generation and management. The law also creates legal requirements to ban specific waste and set a regulation on waste collection, treatment, transportation, storage and recovery. It also advocates aligning state and local government procurement with the circular economy vision, thereby strengthening markets for innovative products and services. In addition, the National Circular Economy Strategy, expected for 2024, will aim to reduce primary raw material consumption to improve market conditions for secondary materials and increase their durability, repairability and circularity. The foundations on which the future strategy will be built recognise SMEs as key actors in circularity and drivers of innovation. A first step to support industry and SMEs will require removing unnecessary regulatory barriers and establishing complementary frameworks. This includes R&D funding for innovation, advisory programmes on resource-efficient production and further development of norms and standards. To foster sustainable consumption, the strategy is expected to develop measures to provide a suitable legal framework, and create the right incentives for companies and consumers through measures on green product design, providing information (for example labelling) or empowering consumers regarding product repairability (right to repair) (Federal Environment Ministry, 2023[83]).

International examples also show that public procurement generally does not incentivise the use of circular business models. The authorities awarding the contracts tend to only consider upfront costs rather than lifecycle costs (e.g. operation, maintenance and end-of-life), and the price is often the dominating criterion for awarding procurement contracts. Meanwhile, when environmental criteria are added to the selection process, in practice, the price still remains the prevailing selection method. Moreover, when introducing environmental criteria, there is a risk for tenders to go empty or that participating companies would complain about the possible threat of anti-rivalry clauses, claiming that only big companies can meet some of the specific requirements. For instance, in Ljubljana (Slovenia), legal constraints on public procurement pose challenges to the emergence and adoption of innovative projects (OECD, 2020[84]). Other regulatory barriers can inhibit the development and implementation of circular economy strategies. In Umeå (Sweden), regulatory barriers are related to the definition of waste (each material is considered waste once it has been collected), which hinders the reuse of some materials because they could generate environmental and health issues (OECD, 2020[13]).

Businesses are often not sufficiently aware of what a circular economy is and how to take advantage of the opportunities it offers. There is an overall lack of understanding of the potential benefits of the circular economy and little interest from companies. The initiatives described in this chapter show that most of the companies involved in the circular economy transition in Hamburg are committed to waste reduction, but few are taking a fully circular approach. Overall, key private sector stakeholders in Hamburg are not informed about the potential economic benefits of circular business models. In some cases, this leads to a perception among businesses that the adoption of circular economy principles entails significant costs and limited prospects for returns on investment. International experience also illustrates this lack of awareness in the private sector. For instance, in Valladolid (Spain), more than 70% of 70 companies surveyed in 2018 declared that they did not know what the circular economy meant. They associated the term with minimising waste production, recycling and reuse, and assumed that they already implemented these processes on a regular basis (OECD, 2020[84]). Similarly, in Ireland, only 51% of businesses understood the meaning of the circular economy (OECD, 2022[85]). Moreover, in some cases, there is a form of scepticism across stakeholders who implement environmental and sustainable practices but do not see the value added of the circular economy approach (OECD, 2022[85]). The lack of cost-benefit analysis of various activities and sectors slows down the transition towards the circular economy. Limited awareness of circular economy practices and their effects amongst key players can hinder opportunities for their implementation and scaling up (OECD, 2020[10]).

There are also insufficient financial incentives to promote the adoption of circular business models. Shifting from a linear to a circular economy presents financial risks for economic actors. Risks can be associated with the critical mass of activities taking place in cities of different sizes, or dimensions related to market size, population, material flows, etc. 10% of businesses responding to the HCC survey have identified the lack of critical mass as well as the lack of financial resources as key obstacles hindering the implementation of circular business models (Figure 4.6). Still, virgin materials are usually less expensive than secondary products, despite the environmental impacts of the former, and uncertainties about the economic benefits can derail an effective implementation of the circular economy (OECD, 2020[10]). For example, in Groningen (The Netherlands), the lack of financial resources for innovators results in small-scale, low-risk projects with limited impact in terms of job creation and positive environmental effects. Local entrepreneurs face high investment risks and maintenance costs (e.g. the cost of secondary materials compared to virgin materials), and when funding is available from innovation programmes for a circular economy, they often lack the knowledge, skills, resources and time to apply for calls (OECD, 2022[85]).

Circular business models only occupy a peripheral position in most markets. Recycled pulp and paper, metals, and plastics represent small proportions of global material output, while remanufactured industrial and consumer products represent an even smaller share of global manufacturing (OECD, 2020[86]). Hence, companies are often reluctant in changing procedures and forms of financing related to the loss of efficient market reach, compared to the competition that continues to work in a linear logic. Consequently, there are difficulties in scaling up pilot projects and experimentation. In addition, the collaboration between stakeholders along the value chain, both upstream and downstream, is challenging due to prevailing competitive dynamics or limited market interactions among these actors (Ekins, P. et al., 2019[56]).

However, the governance gaps identified in Hamburg are similar to those faced by other subnational governments. Box 4.6 shows the main barriers identified by more than 50 cities and regions in OECD countries and their corresponding five broad categories: funding, regulation, policy, awareness and capacity.

Box 4.6. Governance gaps for the transition towards a circular economy in cities and regions

Results from the OECD synthesis report on (2020[10]) "The Circular Economy in Cities and Regions" show that major obstacles for cities and regions transitioning towards circular economies are not technical but of an economic and governance nature (Figure 4.7). More than a third of the interviewed stakeholders in the OECD survey have identified insufficient financial resources, inadequate regulatory frameworks, financial risks, cultural barriers and the lack of a holistic vision as major obstacles.

Figure 4.7. Main obstacles to the circular economy in surveyed cities and regions

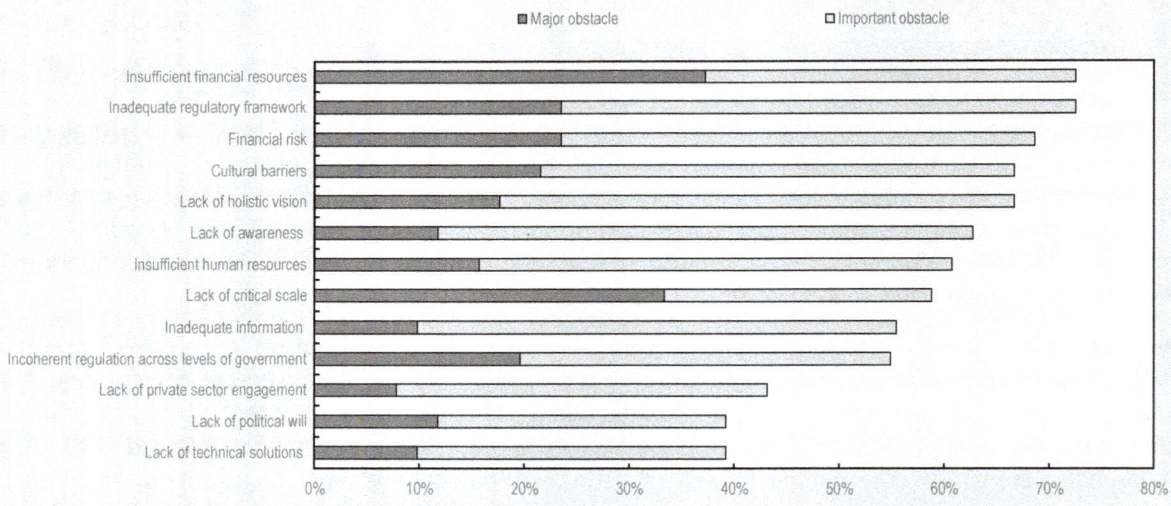

Note: Results based on a sample of 51 respondents who indicated obstacles as being "Major" and "Important"

Source: OECD (2020[10]), *The Circular Economy in Cities and Regions: Synthesis Report, OECD Urban Studies, OECD Publishing, Paris,* https://doi.org/10.1787/10ac6ae4-en.

Key actions to promote circular business models in Hamburg

The following section proposes three types of actions that the HCC can carry out to transition from a linear to a circular economy: (i) immediate actions; (ii) medium-term action that the city of Hamburg can take in co-ordination with the HCC and key stakeholders by 2030; and (iii) targets towards 2040 for the city of Hamburg in collaboration with the HCC.

Immediate actions

As the HCC is a legal advocacy body of the business sector and an intermediary between the local government and the business community in Hamburg, it has an important role to play in fostering a circular economy. In particular and based on international good practices, the HCC could consider implementing the following recommendations as immediate action:

- **Raise awareness of the circular economy and related opportunities to encourage sustainable production and consumption and the adoption of circular principles**. For example, since 2015, the Chamber of Commerce of Glasgow (Scotland, UK), in collaboration with Glasgow City Council and Zero Waste Scotland, has put in place a number of events on the circular economy to raise awareness and facilitate collaborations, workshops and circular economy platforms to crowdsource ideas (e.g. Circle Lab Challenge) (OECD, 2021[14]). London (United Kingdom) has recruited "*circular economy ambassadors*" in different companies and local authorities to share the benefits of the circular economy, with specific information for each economic sector and to raise awareness at the workplace (LWARB, 2017[87]). Collecting data on the progress and the impact of circular economy practices in Hamburg (e.g. food and packaging waste avoided, share of the second-hand textile market) could also help associated companies to better understand the opportunities of the circular economy in practice.
- **Build capacities among the business community.** Through its training subsidiary, the HCC could consider offering training programmes targeting business actors, providing them with deeper knowledge and tools to succeed in their circular projects and discover new business opportunities. Particularly in an intermediary role and in the area of education, the HCC can influence the content of training and lifelong learning programmes, which could be used more intensively toward individual and collective behavioural transformations. For example, in 2018, the Valladolid Chamber of Commerce (Spain) launched a master course in "*Digital transformation and the circular economy*". The curriculum included product life cycle analysis, eco-design, circular value chains and data mining (Valladolid Chamber of Commerce, 2019[88]). The aim of the master course, beyond building specific skills on the circular economy, was also to increase awareness among professionals. The Chamber of Commerce of Valladolid, together with research partners, is creating consulting models for companies willing to adopt circular economy processes (OECD, 2020[84]). In the UK, the Glasgow City Council, Zero Waste Scotland and the Glasgow Chamber of Commerce have collaborated to create the *Circular Glasgow* initiative, which aims to build best practice and capacity with regards to the circular economy, mainly within the private sector in Glasgow. This initiative aims at connecting companies across the city to share ideas and developing circular strategies that can contribute to climate change objectives (Glasgow Chamber of Commerce, 2017[89]). In Sweden, the Umeå branch of the NGO Cradlenet Norr, founded in 2015, organises bi-monthly meetings to discuss challenges with different stakeholders (e.g. SMEs, municipal authorities, business coaches, university researchers and students), organise specialised field visits and participate in international events and platforms on the circular economy. It also provided training courses on circular business models (OECD, 2020[13]).
- **Support the local government of Hamburg through market innovation.** The HCC could collaborate with the city to support market innovation with the creation of incubators, hubs and spaces for experimentation, and it could stimulate demand by being a launch customer. For

example, in Groningen (The Netherlands), a Circular Economy Hub is planned as an incubator space for small businesses and start-ups, and as an information centre, repair hub and second-hand shop (OECD, 2020[90]). The Tallinn Creative Incubator (Estonia) offers on its website seven educational videos on how companies and start-ups can move towards a circular economy (Tallinn Business Incubators, 2022[91]). In Spain, the Business Confederation of Granada created the "OnGranada technological cluster" in 2014, encompassing a diverse membership of private and public sector entities including municipal, provincial, and regional governments, alongside the University of Granada, and the Granada Chamber of Commerce. The cluster is actively integrating circular economy principles into waste reuse and resource efficiency initiatives, such as optimising water usage in plant cultivation and irrigation through injections, as well as exploring opportunities to convert waste from olive production into biofuels (IUC, 2020[92]). The HCC could also build on ongoing initiatives to foster innovation towards a circular economy. For example, the HCC and the City of Hamburg fund the Innovation Contact Service Hamburg (*Innovations Kontakt Stelle Hamburg*), which supports communication between companies and scientific institutions in Hamburg and facilitates mutual access by identifying potential R&D opportunities and networking suitable project partners (IKS Hamburg, 2023[93]). In addition, the HCC's Innovation and Patent Centre supports member companies in the protection of intellectual property rights, patent and innovation management and provides patent-related advice. Services provided by the HCC include awareness-raising events, patent filing assistance, patent potential analysis and market analysis (HCC, 2023[94]).

Actions by 2030

In order to create the enabling conditions for a transition from a linear to a circular economy, the local and national governments, in consultation with the HCC and key stakeholders, could **promote the uptake of circular economy systems through changes in legislation and specific regulatory tools.** These tools can include circular procurement, product norms, industry targets, standards for secondary materials, tax reductions, differentiated fee structures for waste management and investment support (EIB, 2021[95]). For example, since 2020, the city of Turku (Finland) has implemented circular procurement to reduce food waste and GHG emissions from food and related services, in accordance with the Finnish Procurement Act, by doubling the share of vegetarian meals and reducing heating and electricity consumption (City of Turku, 2020[96]). In 2009, Paris (France) developed a Sustainable Food Plan to encourage the procurement of seasonal and local food to boost the local economy and reduce environmental impact. The plan covers 1 200 municipal refectories, including in schools, retirement homes and staff lunchrooms, serving over 30 million meals a year (City of Paris, 2015[97]).

Moreover, it is important to ensure adequate incentives to stimulate large scale and high impact projects, by broadening the range of financial instruments supporting businesses towards the circular economy (from grants to venture capital) and by creating spaces for experimentation. Since 2010 in Paris, France, the Urban Lab has accompanied more than 200 experiments and consolidated a methodology to support effective experimentation in 4 main stages: i) the definition of the experimental project and its evaluation; ii) a search of the experimental site; iii) the deployment of experimentation; and iv) valuation and transformation. In order to facilitate access to these experimental sites, the Urban Lab has been working, for 10 years, in the development of a legal framework that start-ups can refer to for the development of their projects (e.g. a model agreement for using publicly owned spaces for a fixed period of time. Concerning the funding, between 2017 and 2018, the city of Valladolid (Spain) operated a grant programme for circular projects to support the development of local circular initiatives to create jobs and economic prosperity. Through this programme, the local government financed 61 projects for a total budget of EUR 960 000, benefitting in particular private companies, business associations, non-profit entities and research centres based in the city. However, projects struggled to scale up after the experimentation phase (OECD, 2020[84]). In 2022, the city of Montreal (Canada) launched the Open Innovation Grant (*subvention* *à l'innovation ouverte*), which encourages emerging companies to collaborate with established organisations in the city to test innovative solutions in a business context, especially in relation to the circular economy (OECD, 2022[98]). In 2016, the city of Amsterdam (The Netherlands) created a revolving sustainability fund for businesses to pay back within 15 years with a very low interest rate (OECD, 2020[13]).

Lastly, the HCC could act as a **one-stop-shop** for member companies seeking information on circular business models and on regulation and legislation. To this end, the Chamber could create a thematic committee on the circular economy, which could offer information and administrative support regarding circular economy projects for businesses, help reduce transaction costs for entrepreneurs and SMEs, and identify the main regulatory barriers. For instance, the initiative Start-up Slovenia, established in 2014, mobilises a network of mentors from various backgrounds to provide entrepreneurs and young firms with tailored advice. Nowadays, some start-ups also work within the circular economy field (OECD, 2020[10]).

Based on the OECD Checklist for Action for the transition to a circular economy (2020[10]), the City of Hamburg in collaboration with the HCC could meet the targets of the 12 key governance dimensions that would enable a circular economy system to thrive by 2040 (Box.4.7).

Box.4.7. The OECD Checklist for Action for cities and regions and the OECD Scoreboard on the Governance of the Circular Economy

The OECD Checklist for Action for cities and regions aims to support decision-makers in promoting, facilitating and enabling the transition to the circular economy (Figure 4.8). The Checklist is based on 12 key governance dimensions, which are grouped into three clusters corresponding to the complementary roles of cities and regions as promoters, facilitators and enablers of the circular economy.

The OECD Checklist for Action is accompanied by the OECD Scoreboard on the Governance of the Circular Economy (Figure 4.9) which helps governments identify their level of progress towards the implementation of each of the 12 governance dimensions, namely Newcomer (Planned; In development), In progress (In place, not implemented; In place, partly implemented) and Advanced (In place, functioning; In place, objectives achieved).

Figure 4.8. The governance of the circular economy in cities and regions: A Checklist for Action

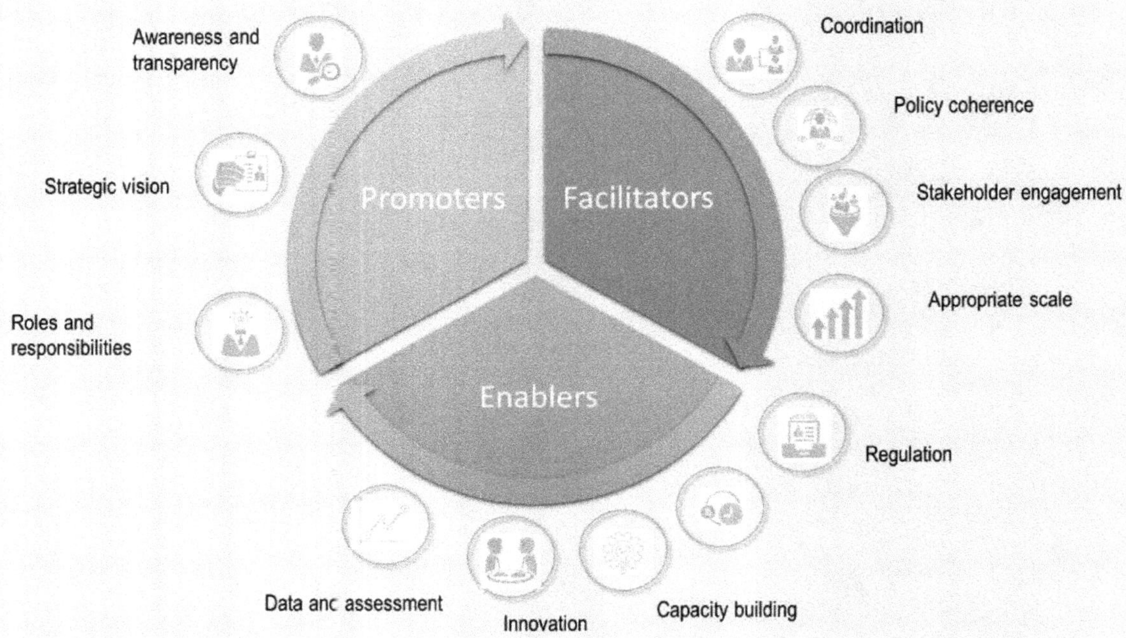

Figure 4.9. Visualisation of the results from the Scoreboard on the Governance of the Circular Economy

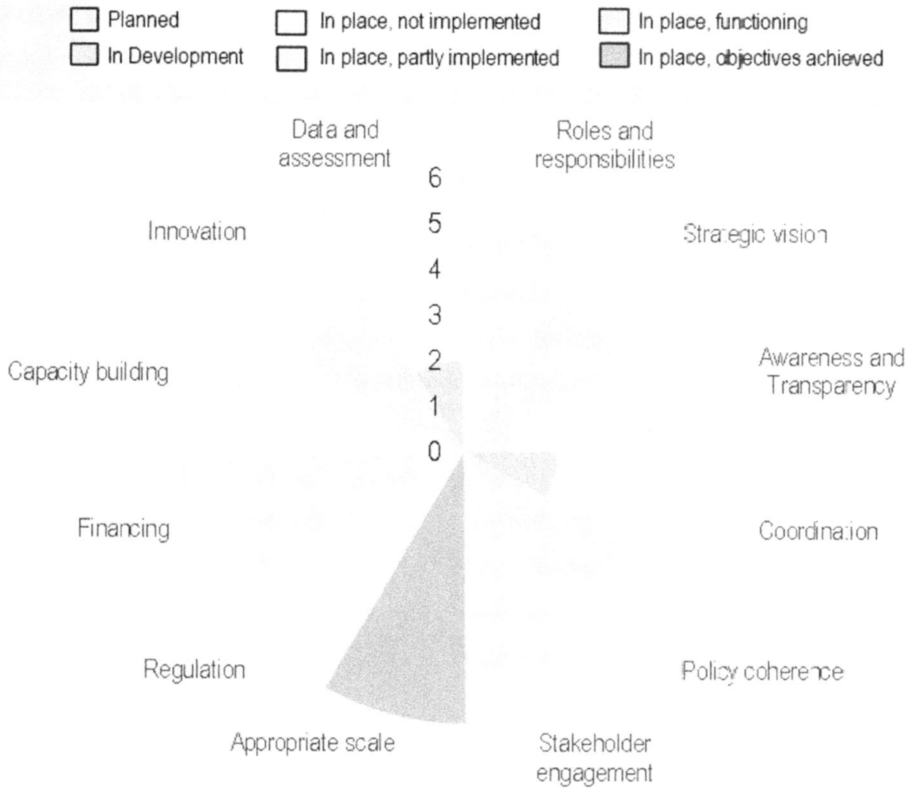

Based on the most "advanced" levels of implementation of the 12 governance dimensions, the city of Hamburg in collaboration with the Hamburg Chamber of Commerce could meet the following targets:

1. **Roles and responsibilities**: The government adopts circular economy principles in all policies and activities. It is a role model for citizens and businesses and leads by example. Roles and responsibilities are clearly allocated across the departments of the city of Hamburg.
2. **Strategic vision**: A circular economy strategy is in place with specific goals consistently achieved and periodically monitored and revised.
3. **Awareness and transparency**: Stakeholders are informed, clear communication is in place. Producers and consumers are aware of the opportunities and the means made available by the government to boost the transition towards a circular economy.
4. **Co-ordination**: Circular economy co-ordination mechanisms across levels of governments are functioning and leading to actions, which are monitored and revised.
5. **Policy coherence**: Departments favour co-ordination and link circular principles to key strategies as business as usual. Grey areas, overlaps, and conflicting objectives are avoided. Results are monitored and checked for further improvement.
6. **Stakeholder engagement**: Stakeholders are actively engaged in the transition towards a circular economy, as its implementation is a shared responsibility. The government facilitates contacts and collaboration.

7. **Appropriate scale**: Circular economy initiatives are embedded in a territorial approach, considering functional rather than administrative boundaries. Results are monitored and follow-up initiatives are considered.
8. **Regulation**: Regulation is fit to foster the circular economy transition (e.g. Masterplan for Climate Protection in Hamburg, Hamburg Climate Protection Law, Waste Management Ordinance). A dialogue across levels of government is established when responsibilities are shared. Results are monitored and initiatives scaled up.
9. **Financing**: Financial instruments are well functioning and impacts are monitored (e.g. promoting systematic recognition of good practices through project audits).
10. **Capacity building**: Specific capacity-building programmes are implemented. They contribute to creating new skills, technical competencies and new jobs opportunities.
11. **Innovation**: The enabling environment for supporting circular business is in place (e.g. regulation, funds) and functioning. The local or regional government provides additional tools, such as co-creation spaces, networks, one-stop-shop for businesses and capacity-building programme.
12. **Data and assessment**: Data is systematically used and updated to inform the public policy design and implementation and to promote circular business models.

Source: OECD (2020[10]), *The Circular Economy in Cities and Regions: Synthesis Report, OECD Urban Studies, OECD Publishing, Paris, https://doi.org/10.1787/10ac6ae4-en*.

References

Blomsma, F. and G. Brennan (2017), "The Emergence of Circular Economy: A New Framing Around Prolonging Resource Productivity", *Journal of Industrial Ecology*, Vol. 21/3, pp. 603-614, https://doi.org/10.1111/jiec.12603. [3]

BMUV (2023), *End-of-life Vehicle Ordinance*, https://www.bmuv.de/en/topics/water-management/circular-economy-overview/overview-types-of-waste-and-waste-flows/end-of-life-vehicles/legislation-in-germany-end-of-life-vehicle-ordinance#:~:text=The%20End%2Dof%2Dlife%20Vehicle,disposal%20compatible%20with. [26]

BMUV (2023), *Substitute Building Materials Ordinance*, http://www.bmuv.de/faqs/mantelverordnung. [33]

BMUV (2023), *Waste Wood Ordinance*, http://www.bmuv.de/en/topics/water-management/circular-economy-overview/overview-types-of-waste-and-waste-flows/waste-wood. [31]

BMUV (2022), *Directive 2012/19/EU on waste electrical and electronic equipment*, http://www.bmuv.de/gesetz/richtlinie-2012-19-eu-ueber-elektro-und-elektronik-altgeraete. [30]

BMUV (2021), *Batteries Act*, http://www.bmuv.de/en/law/batteries-act. [29]

BMUV (2017), *Sewage Sludge Ordinance*, http://www.bmuv.de/en/law/sewage-sludge-ordinance. [32]

BMUV (2002), *Waste Oil Ordinance*, http://www.bmuv.de/en/law/waste-oil-ordinance. [27]

Bridge&Tunnel (2023), *Sustainable design*, https://bridgeandtunnel.de/. [69]

CIRAIG (2015), *Circular economy: A critical literature review of concepts*. [4]

CIRCuIT (2023), *Circular Construction in Regenerative Cities*, http://www.circuit-project.eu/. [52]

Circular Cities Declaration (2023), *Current signatories*. [15]

Cirplus (2023), *Procure Recycled Plastics Online*, https://www.cirplus.com/en. [73]

City of Espoo (2013), *Sustainable forms of transport*, https://static.espoo.fi/cdn/ff/OzEHVL5d23AoPr2wGU1zRuRpspGa0vE__JdfZq9GJ2U/1630052856/public/2021-08/Py%C3%B6r%C3%A4ilyn%20edist%C3%A4misohjelma%202013-2024%2C%20ei%20saavutettava.pdf. [81]

City of Hamburg (2023), *Hamburg eco-partnership*, http://www.hamburg.de/umweltpartnerschaft/. [35]

City of Hamburg (2023), *Shaping a sustainable circular economy together: "circular hub nord" starts with companies and actors*, https://www.hamburg.de/pressearchiv-fhh/17000544/2023-03-22-bukea-kreislaufwirtschaft/. [36]

City of Hamburg (2023), *Sustainability Report of the City of Hamburg*, https://www.hamburg.de/bukea/17216566/2023-06-27-bukea-nachhaltigkeitsbericht/. [39]

City of Hamburg (2023), *What is important about food sharing*, https://www.hamburg.de/lebensmittel/16188464/foodsharing/. [44]

City of Hamburg (2022), *Food in reusable food for Hamburg*, https://www.hamburg.de/einfachmehrweg/15850824/einfachmehrweg/. [34]

City of Hamburg (2022), *Launch of the "catched" initiative: Working together against food waste*, https://www.hamburg.de/pressearchiv-fhh/16338090/2022-07-13-bjv-start-der-initiative-aufgefangen/. [37]

City of Hamburg (2020), *Confident, solidarity-based, sustainable - Shaping Hamburg's future powerfully*, https://www.hamburg.de/senatsthemen/koalitionsvertrag/. [25]

City of Hamburg (2020), *Hamburg Climate Protection Act - HmbKliSchG*, https://www.landesrecht-hamburg.de/bsha/document/jlr-KlimaSchGHA2020rahmen/part/R. [17]

City of Hamburg (2020), *Hamburg Climate Protection Law*, https://www.hamburg.de/klimaschutzgesetz/. [24]

City of Hamburg (2018), *Hamburg drinks reusable to go*, https://www.hamburg.de/pressearchiv-fhh/11864700/2018-11-16-bue-mehrweg-to-go/. [76]

City of Hamburg (2017), *Waste Management Ordinance*, https://www.hamburg.com/publicservice/info/11927336/. [23]

City of Hamburg (2013), *Masterplan for Climate Protection in Hamburg*, https://www.hamburg.de/contentblob/4316146/23474f900f9bf2c0384158f5ee599e03/data/master-plan-for-climate-protection.pdf. [22]

City of Hamburg (2011), *Hamburg Climate Action Plan 2007-2012. Update 2011-2012*, https://www.hamburg.de/contentblob/3959638/a794d37d1560dea335eec5febdf94eae/data/download-update-2011-12-englisch.pdf. [21]

City of Hamburg (2009), *Recycling-Offensive für den Klimaschutz*, https://www.hamburg.de/contentblob/1587654/41df82a4395de075587174124aaebe4c/data/recycling-offensive-klimaschutz.pdf. [19]

City of Hamburg (2005), *Hamburg Waste Disposal Act*, https://www.landesrecht-hamburg.de/bsha/document/jlr-AbfWGHA2005pG2. [18]

City of Paris (2015), *Paris Sustainable Food Plan 2015-2020*, https://cdn.paris.fr/paris/2019/07/24/725249f7bb31e255a00a85dfa98037e6.pdf. [97]

City of Turku (2020), *Supporting circular food systems through public procurement*, https://www.turku.fi/sites/default/files/atoms/files/circular_turku_-_case_study_5.pdf. [96]

Difu (2023), *Circular City*, https://difu.de/projekte/kreislaufstadt#:~:text=Mit%20dem%20Projekt%20%E2%80%9EKreislaufstadt%20%E2%80%93%20Chancen,Kreislaufwirtschaftsst%C3%A4dte%20und%20%2Dinitiativen%20eine%20eigene. [42]

EC (2022), *Cities Cooperating for Circular Economy*, https://cordis.europa.eu/project/id/689157. [51]

EIB (2021), *The 15 Circular Steps for Cities – Second Edition, European Investment Bank*, https://www.eib.org/attachments/thematic/circular_economy_15_steps_for_cities_second_edition_en.pdf. [95]

Ekins, P. et al. (2019), *The Circular Economy: What, Why, How and Where*, https://www.oecd.org/cfe/regionaldevelopment/Ekins-2019-Circular-Economy-What-Why-How-Where.pdf. [56]

Ellen MacArthur Foundation (2021), *Completing The Picture: How the circular economy tackles climate change*, https://ellenmacarthurfoundation.org/completing-the-picture. [2]

EMF (2013), *The circular economy in detail*, https://archive.ellenmacarthurfoundation.org/explore/the-circular-economy-in-detail. [8]

European Commission (2015), *Circular Economy – Overview*, https://ec.europa.eu/eurostat/web/circular-economy. [7]

Eurostat (2022), *Persons employed in circular economy sectors*, https://ec.europa.eu/eurostat/databrowser/view/CEI_CIE011/default/table?lang=en. [59]

Eurostat (2022), *Private investment and gross added value related to circular economy sectors*, https://ec.europa.eu/eurostat/databrowser/view/CEI_CIE012/default/table?lang=en. [58]

Fab City Hamburg (2022), *Fab City Hamburg project*, https://www.fabcity.hamburg/en/fabcity/association/. [41]

Federal Environment Ministry (2023), *Fundamentals of the National Circular Economy Strategy (NKWS)*, https://www.bmuv.de/en/download/fundamentals-of-the-national-circular-economy-strategy-nkws. [83]

Federal Ministry for the Environment, Nature Conservation, Nuclear Safety and Consumer Protecction (2012), *Circular Economy and Safeguard the Environmentally Compatible Management of Waste*, https://www.bmuv.de/en/law/circular-economy-and-safeguard-the-environmentally-compatible-management-of-waste. [82]

Federal Ministry of Justice (2012), *Law to promote the circular economy and ensure the environmentally friendly management of waste (Circular Economy Act - KrWG)*, https://www.gesetze-im-internet.de/krwg/KrWG.pdf. [20]

German Retail Association (2022), *Packaging and reusable packaging*, https://www.hde-klimaschutzoffensive.de/de/infothek/verpackung-und-mehrweg. [78]

Glasgow Chamber of Commerce (2017), , https://www.glasgowchamberofcommerce.com/projects/circular-glasgow/. [89]

Government of Spain (2020), *Spanish Circular Economy Strategy*, https://www.miteco.gob.es/es/calidad-y-evaluacion-ambiental/temas/economia-circular/estrategia/. [11]

Hamburg CIty (2017), *Commercial Waste Ordinance*, http://www.hamburg.de/contentblob/9164258/da4a0f1daf4673becd5f02b4c2e90c15/data/d-gewabfv.pdf. [28]

Hamburg City (2023), *Port Development Plan 2040*, https://hafen2040.hamburg/. [40]

Hamburg City (2023), *Repairing instead of throwing away*, https://www.hamburg.de/stadtleben/4601214/repair-cafes-hamburg/. [79]

Hamburg City (2022), *Save costs, protect resources, avoid waste*, https://www.hamburg.de/newsletter-umweltpartnerschaft-2022-12/16750292/hobenkoeoek/. [77]

Hamburg City (2019), *Horizon 2020 CIRCuIT*, http://www.hamburg.de/international/horizon/13222650/circuit/. [53]

Hamburg City (2016), *Horizon 2020 FORCE*, http://www.hamburg.de/international/horizon/9878642/force/. [50]

Hamburg City (2016), *Horizon 2020 REPAiR*, http://www.hamburg.de/international/horizon/9878644/repair/. [54]

Hamburg Consumer Advice Centre (2022), *Where to put the old clothes?*, https://www.vzhh.de/themen/umwelt-nachhaltigkeit/kleidung-textilien/wohin-den-alten-klamotten. [43]

Hamburg Tourism Office (2023), *Develop Hamburg's tourism sustainably, holistically and together*, https://www.hamburg-tourism.de/das-ist-hamburg/nachhaltigkeit-erleben/nachhaltigkeit-in-hamburg/#:~:text=Die%20Hamburg%20Tourismus%20GmbH%20ist,Tourismusentwicklung%20mit%20gutem%20Beispiel%20vorangehen. [47]

Hamburg Tourism Office (2023), *Discover Hamburg in an environmentally conscious and climate-friendly way*, https://www.hamburg-tourism.de/magazin/10-tipps-fuer-einen-nachhaltigen-hamburg-urlaub/. [46]

Hamburg Tourism Office (2023), *Green Fashion Tours*, https://www.hamburg-tourism.de/sehen-erleben/rundfahrten-fuehrungen/stadtfuehrungen/green-fashion-tours/. [38]

Hamburg Tourismus (2023), *Developing Hamburg's tourism sustainably, holistically and together*, https://www.hamburg-tourism.de/das-ist-hamburg/nachhaltigkeit-erleben/nachhaltigkeit-in-hamburg/#:~:text=Die%20Hamburg%20Tourismus%20GmbH%20ist,Tourismusentwicklung%20mit%20gutem%20Beispiel%20vorangehen. [45]

HCC (2023), *The Innovation and Patents Centre*, http://www.ihk.de/hamburg/en/produktmarken/innovation/innovation-patents-centre-1147560. [94]

Homrich, A. et al. (2018), "The circular economy umbrella: Trends and gaps on integrating pathways", *Journal of Cleaner Production*, Vol. 175, pp. 525-543, https://doi.org/10.1016/j.jclepro.2017.11.064. [5]

IFB Hamburg (2023), *PROFI environment and PROFI environment transfer*, https://www.ifbhh.de/foerderprogramm/profi-umwelt-und-profi-umwelt-transfer. [49]

IFB Hamburg (2023), *UfR - companies for resource protection*, https://www.ifbhh.de/foerderprogramm/ufr-unternehmen-fuer-ressourcenschutz. [48]

IKS Hamburg (2023), *Innovation Contact Service Hamburg*, https://iks-hamburg.de/. [93]

Interfacer (2023), *INTERFACER: Building the digital infrastructure for Fab Cities*, http://www.interfacerproject.eu. [99]

Interreg Europe (2021), *"Hamburg Bottle" - Introducing a closed loop of plastics*, https://www.interregeurope.eu/good-practices/hamburg-bottle-introducing-a-closed-loop-of-plastics. [72]

Irish Manufacturing Research (2020), *CIRCULÉIRE – The National Platform for Circular Manufacturing*. [67]

ISO (2021), *ISO Technical Committee 323*, https://unece.org/sites/default/files/2021-11/2_2_ENG_2021%2011%20ISO%20TC%20323%20presentation_0.pdf. [9]

IUC (2020), *Smart Cities and Blockchain Innovations — IUC Pilot Project Granada Competitive Fund — Webinar 1*, https://www.iuc-asia.eu/2020/10/smart-cities-and-blockchain-innovations-iuc-pilot-project-granada-competitive-fund-webinar-1/. [92]

Lahti, T., J. Wincent and V. Parida (2018), "A Definition and Theoretical Review of the Circular Economy, Value Creation, and Sustainable Business Models: Where Are We Now and Where Should Research Move in the Future?", *Sustainability*, Vol. 10/8, p. 2799, https://doi.org/10.3390/su10082799. [57]

LWARB (2017), *London's circular economy route map, London Waste and Recycling Board*, https://relondon.gov.uk/wp-content/uploads/2021/02/LWARB-ReLondons-circular-economy-route-map-for-London.pdf. [87]

Municipality of Umeå (2019), *Tomtebo Beach - A New Neighborhood with People and Sustainability in Focus*, http://www.umea.se/umeakommun/byggaboochmiljo/ (accessed on 28 January 2020). [65]

National Climate Initiative (2023), *New Database: Energy Consultancy for Retailers*, https://www.klimaschutz.de/de/service/meldungen/neue-datenbank-energieberatung-fuer-den-einzelhandel. [62]

Network for Sustainable Construction and Real Estate Management in Cold Climates (2013), *Umeå. More Sustainable Buildings*, http://hallbarahus.se/wp-content/uploads/2019/02/NHB_Broschyr_April_2013.pdf (accessed on 5 March 2020). [64]

NONOI (2023), *For us, upcycling is not a trend, but an attitude*, https://nonoi-studio.de/pages/slow-fashion. [70]

North Sweden Cleantech (2019), *About Us*, https://northswedencleantech.se/en/about-us/ (accessed on 28 January 2020). [66]

OECD (2022), *Draft - Monitoring progress towards a resource efficient and circular economy*, https://one.oecd.org/document/ENV/EPOC/WPRPW/WPEI(2022)2/en/pdf. [6]

OECD (2022), *OECD Tourism Trends and Policies 2022*, OECD Publishing, Paris, https://doi.org/10.1787/a8dd3019-en. [61]

OECD (2022), *The Circular Economy in Ireland*, OECD Urban Studies, OECD Publishing, Paris, https://doi.org/10.1787/7d25e0bb-en. [85]

OECD (2022), *Vers une stratégie d'économie circulaire à Montréal*, https://www.oecd.org/cfe/cities/Montreal_economie_circulaire.pdf. [98]

OECD (2021), *The Circular Economy in Glasgow, United Kingdom*, OECD Urban Studies, OECD Publishing, Paris, https://doi.org/10.1787/7717a310-en. [14]

OECD (2021), *The Circular Economy in Granada, Spain*, https://www.oecd-ilibrary.org/urban-rural-and-regional-development/the-circular-economy-in-granada-spain_5f8bd827-en. [75]

OECD (2020), "COVID-19 and the retail sector: impact and policy responses", *OECD Policy Responses to Coronavirus (COVID-19)*, OECD Publishing, Paris, https://doi.org/10.1787/371d7599-en. [60]

OECD (2020), *Decarbonising Urban Mobility with Land Use and Transport Policies: The Case of Auckland, New Zealand*, https://www.oecd.org/publications/decarbonising-urban-mobility-with-land-use-and-transport-policies-the-case-of-auckland-new-zealand-095848a3-en.htm. [80]

OECD (2020), *OECD - Nordic Innovation webinars on the circular economy in cities and regions*, https://www.oecd.org/cfe/regionaldevelopment/highlights-2nd-OECD-roundtable-circular-economy.pdf?_ga=2.65296494.1772013035.1686788605-2032697422.1673952863. [86]

OECD (2020), *The Circular Economy in Cities and Regions: Synthesis Report*, OECD Urban Studies, OECD Publishing, Paris, https://doi.org/10.1787/10ac6ae4-en. [10]

OECD (2020), *The Circular Economy in Groningen, the Netherlands*, https://www.oecd-ilibrary.org/urban-rural-and-regional-development/the-circular-economy-in-groningen-the-netherlands_e53348d4-en. [90]

OECD (2020), *The Circular Economy in Umeå, Sweden*, OECD Urban Studies, OECD Publishing, Paris, https://doi.org/10.1787/4ec5dbcd-en. [13]

OECD (2020), *The Circular Economy in Valladolid, Spain*, https://www.oecd.org/fr/espagne/the-circular-economy-in-valladolid-spain-95b1d56e-en.htm. [84]

OECD (2019), *Business Models for the Circular Economy: Opportunities and Challenges for Policy*, https://www.oecd.org/environment/business-models-for-the-circular-economy-g2g9dd62-en.htm. [63]

OECD (2019), *Global Material Resources Outlook to 2060: Economic Drivers and Environmental Consequences*, OECD Publishing, Paris, https://doi.org/10.1787/9789264307452-en. [1]

Recyclehero (2023), *Your pick-up service for waste glass, waste, paper, deposit and old clothes*, https://www.recyclehero.de/. [74]

REPAiR (2023), *PULL Hamburg-Altona & County of Pinneberg (D)*, https://h2020repair.eu/case-studies/hamburg-d/. [55]

Scottish Government (2016), *Making Things Last: A Circular Economy Strategy for Scotland*, https://www.gov.scot/publications/making-things-last-circular-economy-strategy-scotland/pages/4/. [12]

Senate Department for Mobility, Transport, Climate Protection and Environment (2023), . [16]

Tallinn Business Incubators (2022), *Circular Economy*, https://inkubaator.tallinn.ee/ringmajandus/. [91]

Traid (2018), *The Impacts of Clothing*, https://traid.org.uk/wp-content/uploads/2018/09/impacts_of_clothing_factsheet_23percent.pdf. [68]

TUI Cruises (2019), *Powerful impetus for the entire cruise sector*, https://www.tuigroup.com/en-en/media/press-releases/2019/2019-07-18-tc-impetus-for-entire-cruise-sector. [71]

Valladolid Chamber of Commerce (2019), *Masters and Postgraduate Courses 2018-2019*, http://www.escueladenegocio.com/eden/wp-content/uploads/2018/08/Cat%C3%A1logo-Masters-y-Cursos-de-Postgrado-2018-2019.pdf (accessed on 15 June 2023). [88]

Notes

[1] The OECD Survey on the Circular Economy in Cities and Regions was launched in April 2019, and submitted to more than 100 cities from OECD and non-OECD countries. By July 2020, a total of 51 responses were provided by 47 cities, 1 regional county municipality and 3 regions, located in Europe (38), North and South America (10), Oceania (2) and Asia (1).

[2] https://eur-lex.europa.eu/eli/dir/2019/904/oj

[3] The city of Hamburg and the European Union jointly fund the Interfacer project which amongst others, aims to provide Digital Product Passports (DPP) to track and trace material flows (Interfacer, 2023[99]).

[4] The section below provides an overview of the main initiatives already in place in the city of Hamburg, based on the information obtained during three interviews (29 August, 22 September, 17 October 2023) with stakeholders associated with the retail and hospitality industry, as well as desk research.

www.ingramcontent.com/pod-product-compliance
Lightning Source LLC
Chambersburg PA
CBHW061210240926
51846CB00006B/1179
* 9 7 8 9 2 6 4 7 7 2 0 2 1 *